*The Chicago Guide to
Communicating Science*

Chicago Guides
to *Writing*, Editing
and Publishing

The Chicago Guide to Communicating Science

SECOND EDITION

SCOTT L. MONTGOMERY

The University of Chicago Press
Chicago and London

The University of Chicago Press, Chicago 60637
The University of Chicago Press, Ltd., London
© 2003, 2017 by The University of Chicago

Published 2017
Printed in the United States of America

26 25 24 23 22 21 20 19   2 3 4 5

ISBN-13: 978-0-226-14450-4 (paper)
ISBN-13: 978-0-226-14464-1 (e-book)
DOI: 10.7208/chicago/9780226144641.001.0001

Library of Congress Cataloging-in-Publication Data
Names: Montgomery, Scott L., author.
Title: The Chicago guide to communicating science / Scott L. Montgomery.
Other titles: Chicago guides to writing, editing, and publishing.
Description: Second edition. | Chicago ; London : The University of Chicago Press,
2017. | Series: Chicago guides to writing, editing, and publishing.
Identifiers: LCCN 2016025836 | ISBN 9780226144504 (pbk. : alk. paper) |
ISBN 9780226144641 (e-book)
Subjects: LSCH : Communication of technical information. | Communication in
science. | Technical writing.
Classification: LCC T10.5.M65 2017 | DDC 501/.4—dc23 LC record aailable at
https://lccn.loc.gov/2016025836

To Kay and Shirley, once more and always

CONTENTS

PREFACE TO THE SECOND EDITION

A new edition of any work provides the author with both a compliment and a second chance. Readers have found the first version worthwhile but now want it improved. This demand is a good thing. In such a rapidly changing field as scientific communication, one is able to update critical material, correct errors, sharpen important points, remove embarrassments, and even take account of the comments (always kindly intended, to be sure) of reviewers and other commentators. No less, the author is able to expand and deepen a book's material so that it more fully matches the historical changes that have happened.

So it is in this case. Since the first edition of this book appeared, the communicational world of science has altered itself profoundly. It seems no exaggeration to say that, since the late 1990s, this world has entered a new epoch. The Internet now clearly stands at the center of science and brings ever-new opportunities for expressing, sharing, and abusing knowledge, opportunities as exciting and productive as they are confusing and unresolved. The overall effect, however, is that demand for communicational skill on the part of scientists and uses for this skill are greater than ever.

This second edition responds directly to this state of affairs. As before, my aim has been to provide wholly practical advice in the context of increasing a reader's awareness of why such advice makes sense. At the same time, I have extended such advice into a number of new areas. Today, more than ever before, scientists are being called upon to address an array of audiences. Understanding the spectrum of skill required to do this competently and excellently qualifies as an essential way to keep the image of science itself strong and vital.

New material in this edition includes chapters on plagiarism and fraud, writing a graduate thesis, science translation, writing and speaking to nontechnical audiences, and teaching science communication in the classroom. Selected enhancements have been given to other portions of the book, such as the chapters on the language of science, on writing very well, and on interacting with the press. On the other hand, I have kept the figures for discussion from the first edition. Though Internet capabilities have certainly expanded the visual dimension to science, the fundamental points related to creating superior images haven't changed.

The book also has a revised structure. It is now divided into three parts: the first deals with the language and rhetoric of science and how this works in good writing, the second part concerns the specific forms of professional communication for scientists and how to generate them, and the third part deals with special topics, including most of the new chapter subjects noted above.

My hope and belief are that this new edition proves I have taken good advantage of a second chance and provided readers with a worthy successor to an earlier effort.

Acknowledgments are needed. Rather than expressing a debt, however, I prefer to offer compliments. Mary Laur, of the University of Chicago Press, has been a wonderful help during the creation of this edition: engaged, understanding, and, of course, patient. I can only hope to work with her again on a future project. With her help, I also had the benefit of two outstanding (anonymous) reviewers, whose comments improved this volume and returned to its author a deserved awareness of his own limits. Some excellent suggestions were also rendered by Christie Henry, editor extraordinaire of the same press. Kelly Finefrock-Creed saved me from some key embarrassments and raised the manuscript to a higher level than it might otherwise have achieved. Many of my students over the years have contributed good ideas and experiences that I have gratefully absorbed and for which I hope I have thanked them sufficiently. Finally, as ever, it is Marilyn, Kyle, Cameron, and Clio who made this book both possible and hopeful. *Nemo est insula, sed peninsula.*

PREFACE TO THE FIRST EDITION

This book is a product of much time and labor spent in the halls, towers, moats, and dungeons of scientific communication. My own work as a publishing scientist and as an independent scholar in the history of science and scientific language has provided much material to draw upon. But this is only part of the story. Like a great majority of writers, I've been both peasant and yeoman in the fields of contemporary authorship. This service has entailed being urged or dragged into many roles: essayist, freelancer, translator, scriptwriter, speaker, critic, reviewer, ghostwriter, copywriter, editor, fix-it boy, messenger, secretary, and (not least) rejectee. These are the many faces of the "successful" writer in contemporary society—the writer, that is, whose work is allowed to see print on something approaching a regular basis. The chapters that follow come directly from such multipronged experience.

This book is quite likely to differ from most other guides to scientific communication you might come across. The reason is that, at base, I treat scientists not as literary underdogs—that is, as scientists *first*, who must also sometimes (somehow, against whatever odds) express themselves intelligibly—but instead as full-fledged writers or speakers, who understand that the transfer of their knowledge to others is part of the essence of research. Writers, in particular, learn to write most of all not from paternal rules and standards, nor from the mother's milk of step-by-step advice. They learn from their brethren, from other writers—the Latin term, *imitatio*, which evokes the combined sense of imitation, adaptation, and invention, is a good one to apply and, in fact, derives from the rhetorical tradition, from ancient to modern times, of teaching composition. Learning from other writers, however, is not as simple as it may sound. There are a number of important and practical aspects to it that I have sought to integrate, at various levels, throughout this book.

The result is a guide that views the scientist, in part, as a species of the genus "writer" (or rather, "communicator"). It has been my intent to offer a series of clear and realistic choices for how to develop skill at either a functional or superior level, on a personal basis, and in direct awareness of the realities of writing, speaking, and publishing in science today. At the same time, part of this book is also aimed at teaching the scientist-author something about the nature and history of his or her own discourse, as a living, evolving phenomenon. Knowledge of the medium, as well as how to wield it effectively, can only give one a degree of added control over it. And indeed, such control is really the crux of the matter.

Noam Chomsky, quite likely the greatest linguist of the past half century, once called the dictionary a "list of hints." His point was that words and meanings are too pliable in their uses to be held prisoner by a fixed set of definitions. Similarly, those who hope to find the eternal, unchanging rules of good scientific expression are bound to drink from many fountains while still growing old. The flow of science, in form and articulation, has always been diverse, offering reasons for writers and speakers to both adapt and explore.

The writing of this book has benefited greatly from help supplied, over time and not always intentionally, by many friends, teachers, and colleagues. I have space here to mention only a few to whom I owe the deepest gratitude. Bill Travers, geologist and friend, formerly of Cornell University, encouraged many useful discussions over the years on science and writing. Nigel Anstey, geophysicist *magnum*, helped reorient my thinking about certain important subjects. During the past five years, I've received the benefit of helpful criticism from the many reviewers and editors at the American Association of Petroleum Geologists, the Geological Society of America, the Kansas Geological Society, the Rocky Mountain Association of Geologists, the Utah Geological Survey, and the US Geological Survey, to name only the most prominent sources of obligation. Useful exchanges and forums for talking about scientific language and publication were provided, along the way, by Steve Fuller, Les Levidow, and Kirk Junker, with often quiet but nonetheless effective support from John Lyne. Tom Cottner, of the University of Washington Medical Research Center, provided valuable material and information for the present volume. I would also like to thank the reviewers of the original manuscript, who, by their intelligence and knowledge, improved many sections of the book and saved me from certain mistakes of information and judgment. Any remaining errors are entirely my own.

Responsibility for making this book a reality and for ensuring that it retains whatever quality it might possess must be given also to Susan Abrams of the University of Chicago Press. Her unflagging support, patience, and intelligent suggestions during all phases of the project will always be remembered.

Finally, I offer this work as homage to family members, living and passed on, especially Kay, Shirley, Lynn, and Frank, all of whom would have expressed surprise to see such a book emerge from these hands. Lastly, I must thank those "within the palisade"—Kyle, Cameron, and Marilyn—who have endured yet another season of hopeful labor.

＊

Part 1. The Language and Rhetoric of Science

USING THEM TO YOUR ADVANTAGE

1. COMMUNICATING SCIENCE

*If one tells the truth, one is sure, sooner or later,
to be found out.* —OSCAR WILDE

First Things

Science exists because scientists are writers and speakers. We know this, if only intuitively, from the very moment we embark on a career in biology, physics, or geology. As a shared form of knowledge, scientific understanding is inseparable from the written and spoken word. There are no boundaries, no walls, between the doing of science and the communication of it; communicating *is* the doing of science. If data falls in the forest, and no one hears or sees or it . . . Research that never sees the dark of print remains either hidden or virtual or nonexistent. Publication and public speaking are how scientific work gains a presence, a shared reality in the world.

These basic truths form a starting point. As scientists, we are scholars too, steeped in learning, study, and, yes, competitive fellowship. Communicating is our life's work—it is what determines our presence and place in the universe of professional endeavor. And so we must accept the duties, as well as the demands and urges (and, fortunately or unfortunately, the responsibilities) of authorship. But aside from noble sentiment, there are other reasons for being able to communicate well with our intellectual brethren.

No one who aspires to a scientific career can afford to overlook the practical implications of what has just been said. The ability to write and speak effectively will determine, in no uncertain terms, the perceived importance and validity of your work. To a large degree, your reputation will rest on your ability to communicate. The reason to improve your skill in this area, therefore, is not to please English teachers past and present (though these may well haunt us till we shed our mortal coil). It is to gain something very real in the professional world, something of advantage. To communicate well is to engage in self-interest. Another way of saying this is that writing and speaking intelligibly are required forms of professional competence—nothing less.

Contrary to what you may feel, however, based on your own experience and the stories of others, this situation is not a fatal one. Creating and sharing knowledge are truly profound but also eminently performable acts. Indeed, they are among the highest achievements of which human beings are capable. Every time you put finger to keyboard, step up to the podium, or clear your throat in front of a class, you become a full participant in what has clearly become humankind's most powerful domain of intellectual enterprise.

The purpose of this guide is to help you, the scientist, deal competently, even eloquently, with your role as an author. My intent is to aid you in learning how to feel at home with, and even take significant pride in, the communicating you will do as a member of the greater scientific community. This can be done, as it happens, without torture or torment, golden rules or iron systems. What it does require, among other things, is patience, a willingness to learn from others, and a certain way of looking at authorship.

The Importance of Attitude

Writing, we know, does not always come easily to scientists. Innumerable tales can be told of brilliant researchers whose papers would blind the eye of a first-year composition instructor. Yet, in reality, good writing rarely comes easily to *anyone*, in *any* discipline, whether quantum mechanics or art history. Writing is aptly called a skill, or, more accurately, a collection of skills. It is never entirely mechanical and always involves a level of emotional engagement, as well as forbearance and discipline. The Japanese have an excellent proverb for what it takes to learn a skill: "Ishi no ue ni, san nen." Three years, standing on a rock.

I'm not suggesting that we try this (one to two years, with time off for good behavior, should be plenty). But it points in a certain direction. What has our training, as scientists, been like in this area? In fact, a major difference between the humanities and sciences is that composing, critiquing, and revising papers forms a central part of learning in the former, while in the sciences it does not. Moreover, immersing oneself in eloquent writing of the past is also prominent in humanities training, whereas scientific instruction tends to avoid this sort of thing almost entirely. We don't read Newton (or much of him) in a basic physics class, Linnaeus in a botany course, Lavoisier or Lyell in a chemistry or geology curriculum. Why is this so? The reasons are complex, and have much to do with the recent history of science. But the effects are clear: good writing is something that scientists are supposed to pick up, either from a course or two in technical writing while in school, or through osmosis after entering the caffeine-ridden world of professional research.

If formal communication can be intimidating for scientists and engineers, what is the best way to help gain back the upper hand? Much begins with how one thinks about writing in particular and about scientific language in general. To communicate well, you need to feel at least some degree of *control* over the language you are using. This means a basic awareness that you, the writer, are taking words and images and creating something out of them. It also means an understanding that you are doing this by employing certain forms and structures toward the goal of persuading— telling a story to—a very particular kind of audience.

Too often in science we have the feeling that language is our opponent, something we have to wrestle with and subdue. Technical speech can seem like something hardened and formal that we have to obey, that predetermines a great deal of what we can and cannot say. There is a drop of truth here; scientific writing *is* generally flat, unromantic, heavily reliant on preexisting technical terms and phrases. Journal editors are unlikely to smile favorably at literary turns of phrase, passionate outbursts, or fanfares to the gods of invention. Yet this hardly describes the whole of the matter. Science may sound anonymous to the ear, but it is fully human and personal to the touch. The calm, declarative "voice" of technical speech is something we must make anew, every time, through a host of choices, a number of which are actually quite flexible. If we look closely enough, we can find many avenues where personal eloquence may be put to practical use. The creative and the individual have a very important dimension in our writing (I'll say more about this in chapter 4).

At the same time, we scientists have certain advantages over our (distant?) cousins in the humanities. Some of the same aspects that make our language seem flat and formal work in our favor. Abundant use of technical words and phrases does, in fact, mean that pieces of our discourse are prefabricated. There are more moments, that is, during the composition of any paper when a series of words flow easily from the fingers into place, as if by automation. This is not a sign of cybernetic rebirth, but actually something close to the opposite: an intuitive sense of when this is needed or possible. How do we acquire this? The answer is probably not very shocking—by internalizing the discourse of our subject and field. Such can come from long years of reading and reciting (at meetings) the relevant literature, until it becomes second speech. But there are other ways that require far less time, that graduate students can use. I will go over them in chapter 3. The point here is that scientists shouldn't feel that writing is a lonely chore or errand in the wilderness. It is communal at every step and comes with help.

Much begins and ends with attitude, therefore. Reasonably confident authors transfer their sense of self to the reader. Their science tends to be effective, less hesitant. If, however, you are terrified of writing, it is likely that your writing will terrify others (or worse, inspire humor). Conversely, if you view the composition of technical papers as an unbounded creative exercise, with enthrallment as its goal, you will meet a quick and scarlet end at the hands of the first editor you encounter This book has been written to protect you from both fates.

The Existing Literature on Technical Communication: A Brief Warning

I would be remiss, both as a scientist and as a writer, if I did not include some pointed words about my competitors. In technical terms, this means a "review of the existing literature."

Many manuals and guides have been written over the years to fill the training gap in scientific writing and speaking. As might be expected, the results are (to put it diplomatically) variable. There are many excellent thoughts scattered through this literature, like glittering jewels in gray sand. But there is also much glass and cinder. Some points of warning are worth mentioning.

To begin with, many books on scientific communication boil down to collections of rules, standards, and warnings. Some even claim to offer the opposite, but end up embracing the enemy. Such books will tell you:

"Keep all your sentences short and simple" and "Avoid emotional terms." They may order you to "employ the active tense whenever possible" or to "follow the IMRAD structure (Introduction, Methods, Results, and Discussion) in all your papers." And so on. This type of advice, if viewed with the rigor of its own prescriptions, becomes a list of absolutes, like Martin Luther's *Ninety-Five Theses*, to be nailed to the door of every science department in the land.

From a certain point of view, the learning of rules makes sense. Science, after all, is awash in protocols, principles, and standards. Why not apply this to writing? Certainly it can be done. But let us be clear about what it means. The real focus is less on writing per se than on obeying codes of authorial behavior. One is not encouraged to be a true apprentice, to learn from the writing of other scientists, but instead to submit and conform to regulations. That is why these manuals so often adopt a tone of law enforcement ("You should never . . ."). But there is a deeper problem. Rule-driven advice can easily overwhelm us and validate any discomfort we already feel toward writing. Tiptoeing through a minefield of potential errors does little to advance confident steps toward the authorial act. Such advice thus tends to provide us more with the measure of our failures than aids to our success.

Let me give a specific example. Many manuals spend much space laying out precise standards for various items—references, tables, format, article structure, and so forth. Most or all of this is likely to be of little or no value. No universal standards exist for such elements. Different fields often handle them differently. This is just as true for journals, even within single fields. For such reasons, studying the literature of your discipline is the *only* guaranteed way to gain practical knowledge of these conventions.

This brings up another problem area. Authors of writing guides in science tend to offer counsel that reflects their own (inevitably limited) experience. What is good for biomedicine or agronomy, however, is not necessarily good (or even relevant) for chemistry or cosmology. The supposedly universal IMRAD structure is nothing of the kind. Appropriate to experimental work, it is rarely, if ever, followed in large portions of the geosciences, mathematics, physics, engineering, and many other domains where fieldwork, theory, and descriptive efforts are on exhibit. There has never been a single standard for scientific papers, and saying there should be is like claiming there is one and only one procedure for performing all experiments. Any attempt to call for universal standards smacks of authoritarianism, in a domain that has long proven adept at resisting all such impositions. Like nature, scientific work is highly diverse. Needed instead of despotic law is kind advice on *how to learn* what is accepted practice.

Such is all the more true since aspects of scientific publication are in flux in the early 21st century. The world of scholarly publishing as a whole has entered a period of dynamism but also uncertainty. Of course I am speaking about the online universe, which is where just about all scientific expression is headed, if it isn't there already. Rules and preferences, as well as required information, for online papers have evolved, with some journals now asking for inserted links to references, "additional" or "supplementary" materials, and more. The rise of open-access journals, in their various forms, continues to change the landscape of scientific publishing in major ways that all professional scientists need to understand. To that end, since the *how* of publication has large impacts on the *what*, this book will devote some important pages to these topics.

The Approach of This Guide

This is a book about professional scientific communication—what it is, how it can be achieved, understood, and improved. It is written by someone with long experience as an author and presenter both of scientific material and scholarly studies on scientific language. During my career as a geoscientist, I have authored an immodest number of technical papers, monographs, reports, and proprietary studies. At the same time, I have long been fascinated with the discourse of science and have written books and scholarly papers on the rhetoric of science, its historical evolution, its character in various languages, and the translation of it. What appears in this book, therefore, comes from both experience and knowledge.

The focus is on written expression. This is what every scientist must know how to do, bar none. There are also chapters on professional speaking, dealing with the press, communicating with the public, and other topics. Though mainly a book for scientists who write for other scientists, it extends its reach into other key areas where scientific work is communicated.

Fair weight is given to the journal article. Though admittedly a small subset of the total range in technical expression, the journal paper is the dominant—and most scientists and institutions believe the most important—form in which scientific knowledge continues to reside. The scientific journal began 350 years ago, became prevalent in the 19th century, and evolved into an inarguable standard during the 20th century. It may change in the future; new forms of exchange may well emerge. But for now, and the foreseeable future, both online and hard copy science will continue their loyalty to journal-type publication.

So what kind of writing *is* scientific writing? There are two answers, both essential. First, scientific writing is storytelling. You will hear this from other writing guides, and they are right. Consider the subheads of a paper (any paper): it is apparent we are being told about something that happened—what it was, how it was made to occur, what resulted from it, and what it means. But there is a second dimension, too. Scientific writing is also engaged in rhetoric—it aims not just to tell but to persuade. It wants to convince us that the result not only has meaning but is *meaningful*. Such is no less important than the story; indeed, it needs to be the point of the story, as we will see.

In general terms, this is a book of advice, not rules; guidance, not demands. It is my experience, from years of publication and teaching, that scholars of any stripe learn best how to write well if they are addressed *as writers*, not as mere laborers, toiling in the mills and quarries of the word.

What does this mean? A certain shift in dignity, to begin with. But more to the point, it means providing you, the writer, with certain understanding, techniques, and attitudes that will aid you in gaining command over the language you produce and consume for a living. This I hope to do in three fundamental ways. First, I review some points on the nature and history of scientific discourse—this gives us context and a realistic sense of what we can expect of ourselves. Second, I maintain that good writing very often has a base in reading—I mean, reading as writers do, with a critical eye and an ear for quality, for what is worthy of imitation. This leads directly to the third and final point: good communicators learn from others, by identifying and studying examples of successful expression in their chosen field.

This last idea is probably the most important of all. It is a very old and deeply tested truth: authors acquire a comfort and facility for writing by first emulating the excellent work of others. This has always been true, and often admitted, for poets, novelists, playwrights, essayists, and scholars generally. Indeed, the use of models was a central aspect of Western education from at least the time of Quintilian (first century BCE) down to the late 19th century (why this changed is a complicated story). As a general method, it remains very much alive in the arts and humanities today. Experience teaches that scientists most often learn to write this way, too, though on a haphazard basis, since we don't tend to acknowledge it very much or make it an overt part of training.

The thoughtful use of positive and negative models, however, has another prominent advantage. It allows you, the writer, to chose your own teachers (or coaches, if you prefer). Writing is a personal activity, as I have said. But it also makes you part of a community of producers, such that you can improve your skills by drawing on the good work that other members have done. I will have more to say on this matter below. For now, let me leave you with a phrase by one of America's preeminent poets, T. S. Eliot, who once suggested that no artist is ever a complete original but must be set "for contrast and comparison, among the dead."

A word concerning what this book is *not* about. It is not about teaching you grammatical rules or proper scientific usage. There are other volumes along these lines; this book assumes that you are able to form a competent sentence in this language, at least some of the time, and that you know how to use a dictionary. If so, read on; this book is for you. It is also for those with English as a foreign language, to whom I have devoted a separate chapter. Those unable to write grammatically in any language need to begin somewhere else, however. They should have been either bored or scared off by what has been said to this point in any case.

Please think of using this book in several different ways. The next three chapters (2–4) form a unit, lay out the major themes, and will be most rewarding if read together (in order, if possible). Chapter 5 takes some of these themes to a higher level and may not be for everyone. Succeeding chapters, on the other hand, can be either perused in similar fashion, one after another, or dipped into, one at a time, as need or interest arises. If nothing else, I would like you to come away from this book with a changed view of scientific expression—what it is, what makes it up, where it is going, and, above all, how to use it. If even part of this is achieved, a good deed will have been done.

A Final Introductory Word: Philosophies of Language

I have said that the way in which one views language has an important effect on how one uses it. Scientists have been prey, for some time, to a particular philosophy of language that tends to derail their understanding of what might be termed "the scientific message." I refer, specifically, to the overriding maxim "Simplify, simplify." There are many variations on this theme; no doubt you've heard some of them: "Use as few words as possible," "Eliminate anything that is not essential," "Scientific writing must be transparent, a mere vehicle," "Use the active voice at all times," and so forth.

All such ideas exhibit a deep misconception about the nature of technical discourse. The "simplify to the nth degree" mentality is a way of declaring martial law on the inevitable complexities of scientific communication. Besides embodying a philosophy of distrust, this way of thinking lacks any appreciation for the rhetorical *flexibility* of technical writing, as a form of human expression, and the range of literary techniques such writing normally includes—indeed, *must* include. To persuade and convince a highly critical audience, authors cannot simply brain-dump information onto paper. If they could, there would certainly be no need for a book of this type. We would all be masters, with no need of apprenticeship.

Let me give an example. One of the rules most common to the "simplify" philosophy is that the scientific writer should do away with any and all phrases such as "under these or similar circumstances," "it is important to note," "for the most part," "it is doubtful that," and so on. These kinds of fragments, however, though perhaps inessential as far as the data goes, perform a required function in good writing. They act as transitions between sentences or paragraphs and serve as helpful cues for the reader. They add pacing, flow, and important internal connection to the argument.

Effective arguments in any area of study, that is, employ a host of persuasive techniques. Many such techniques, in fact, are used equally by scientific and literary writing, though in different ways. This can be easily shown by a close analysis of any technical paper (see chapter 2). At a fundamental level, there is no deep divide between the sciences and the humanities when it comes to the basics of expression. Only, perhaps, a series of guarded trenches.

A main goal of this book is to help make scientific writers and speakers aware of the forms that they are using, or might use, when they produce competent science. This means learning to read with a critical eye and understanding how specialized the scientific message really is. Writing, in particular, is a messy business. It is as full of trial and error, dead ends, frustrated effort, and minor triumphs as any other part of research. What eventually emerges (hopefully) is a reasonably well-organized, logical flow that hides most (but never all) of this struggle. In the words of Peter Medawar, Nobel laureate in medicine and frequent author on matters of science, "the scientific paper is a fraud." But then, so is all successful writing.

Scientific communication is highly stylized—far more stylized, in fact, than forms such as the literary essay. When we look back at the past, say to the 17th century, and trace technical expression forward, we find that what we are doing when we write is telling very condensed, extremely formalized "stories" to an equally particular audience. In most cases, we have learned to do this through imitation, another trial-and-error process. Consciously or otherwise (usually otherwise), we are employing strategies to convince the reader of our knowledge, competence, originality, and contribution. This seems a tall order, when put this way. It is both ordinary and magnificent. Perhaps the sense that all of this is going on helps make us the critical, scrutinizing, and often skeptical beings that we are. But it should also reconnect us with the reasons why we originally chose to do science, the wonder and fascination, the ambitions and desires, that propelled us in this direction. Writing is about these aspects of our lives, too. Scientists are also writers because science is a great presence in the world.

2. THE LANGUAGE OF SCIENCE: HISTORICAL REALITIES FOR READERS AND WRITERS

All that a scientist creates in a fact is the language in
which he enunciates it. —HENRI POINCARÉ

Matters of History

As scientists, we are largely creatures of the contemporary. Unlike other areas of study, our training provides little on the history of our discipline or the development of our discourse. Does scientific language even *have* a real history? What would Newton or Laplace make of a recent article in *Physical Review Letters*? If Darwin's *Origin of Species* were submitted to a major scientific publisher today, what might be its chances of acceptance (or should we say, survival)?

These are not merely academic questions. Scientists communicate with each other in a professional dialect, one that has evolved. Language stands still for no one, and scientific language is no exception. If you doubt this, I urge you to read through articles in your own field written 50, 75, and 100 years ago: the differences from today will be both obvious and subtle. In fact, let me offer a few examples here.

Anno: 1672.   In the year 1666 . . . I procured me a triangular glass prism, to try therewith the celebrated phaenomena of colours. . . . It was at first a very pleasing divertissement, to view the vivid and intense colours produced thereby; but after a while . . . I became surprised to see them in an oblong form; which, according to the received laws of refraction, I expected should have been circular.[1]

1760.   Beside the horizontal division of the earth into strata, these strata are again divided and shattered by many perpendicular fissures, which are in some places few and narrow, but oftentimes many and of considerable width. There are also many instances, where a particular stratum shall have almost no fissures at all, though the strata both above and below it are considerably broken: this happens frequently in clay, probably on account of the softness of it.[2]

1868.   The extent to which a country suffers denudation at the present time is to be measured by the amount of mineral matter removed from its surface and carried into the sea. An attentive examination of this subject is calculated to throw some light on the vexed question of the origin of valleys and also on the value of geological time.[3]

1965.   In arid climates the rocks exposed to the blazing sun become intensely heated, and in consequence a thin outer shell expands and tends to pull away from the cooler layer a few centimeters within. Under perfectly dry conditions, however, the stresses so developed are insufficient to fracture fresh, massive rocks. Experiments leave no doubt about this.[4]

2014.   Ancient saline fracture waters in the Precambrian continental subsurface, with groundwater residence times ranging from millions to billions of years, provide a previously underestimated source of H_2 for the terrestrial deep biosphere. Until now, little of the information on H_2 in these settings, accessed via underground research laboratories and mines, has been incorporated into global geochemical and biogeochemical models.[5]

1.   I. Newton, "New Theory about Light and Colors," in *Isaac Newton's Papers and Letters on Natural Philosophy*, ed. I. B. Cohen (Cambridge, MA: Harvard University Press, 1958), 47. This is often acknowledged to be the first modern scientific paper.

2.   J. Mitchell, "The Earth Composed of Regular and Uniform Strata," in *A Source Book in Geology, 1400–1900*, ed. K. F. Mather and S. L. Mason (Cambridge, MA: Harvard University Press, 1970), 84.

3.   A. Geikie, "On Denudation Now in Progress," in Mather and Mason, *Source Book in Geology*, 523.

4.   A. Holmes, *Principles of Physical Geology* (New York: Wiley, 1965), 248.

5.   B. S. Lollar, T. C. Onstott, G. Lacrampe-Couloume, and C. J. Ballentine, "The Contribution of the Precambrian Continental Lithosphere to Global H_2 Production," *Nature* 516, no. 7531 (2014): 379.

Even in so brief a selection, we are witness to a crucial evolution in scientific expression over the span of the modern era. We begin with Newton, who gives us a personal tour of his experiments and observations; he is there, with us, gesturing in the room. By the second excerpt, he is gone. The passive tense takes over and the object of interest ("strata") performs the action. Phenomena are now the main characters of the story. Because of this, the writing gains a more objective, formal tone. But as we continue toward the present, the writing undergoes a still more potent increase in density and reliance on terminology. By the final passage, nearly all literary or conversational touches are gone; any fluids of informality have been squeezed out. The style seems mechanical, Euclidian, even ceremonial (actually, it is none of these things, as we will see). Such is the direction our discourse has taken. How, then, did this process begin?

Where We Came From: The Beginnings of Modern Scientific Expression

Modern scientific writing in English began in the 17th century, with authors such as Francis Bacon, Robert Boyle, and Isaac Newton. This period was characterized by intense debates over the nature of language generally. At issue was the presumed power of words to control knowledge, as Bacon put it, to "force and overrule the understanding, throw all into confusion, and lead men away into numberless empty controversies and idle fancies." Bacon was thus the first to claim revolt against Elizabethan styles of writing (which, of course, included Shakespearean drama); these, he said, pulled a veil between the intellect and the world. To advance knowledge, especially "the new experimental philosophy," there was needed a simple, direct, and unadorned form of speech. This would lift the veil and provide "an equal number of words as of things."

Bacon's followers took his ideas very much to heart and made them a philosophical nucleus for the new Royal Society of London, the first scientific society in the English-speaking world. How closely did these men adhere to Baconian principles? Thomas Sprat, in his *History of the Royal Society* (first published 1667), gives us some idea:

Who can behold, without indignation, how many mists and uncertainties, these specious tropes and figures have brought on our knowledge? . . . [We of the Society] have therefore been more rigorous in putting in execu-

tion the only remedy that can be found for this extravagance; and that has
been a constant resolution, to reject all the amplifications, digressions, and
swellings of style; to return back to the primitive purity and shortness . . .
[to] a close, naked, natural way of speaking . . . to bring all things as near
the mathematical plainness as they can; and preferring the language of artisans, countrymen, and merchants, before that of wits, or scholars.[6]

Those of the Royal Society were never more flowery than when denounc-
ing the Elizabethans.

Yet a new style did emerge, by the end of the century. The society had
established a journal (the *Transactions*), which mainly published lectures
given during meetings. That the earliest scientific papers in English very often had to be read aloud in front of an audience did, eventually, impose certain changes in style and length. Newton's "New Theory about Light and
Colors" (1672) helped set a standard. The paper was written as a letter to
Henry Oldenburg, then president of the Royal Society, in a form to be read
aloud to the membership. Newton's paper showed how effective it was
to confine one's speech to a demonstration, a repeating in words of what
was done in actions. Newton, meanwhile, had drawn on, and simplified,
the writing of Robert Boyle, who, as it happens, may well have modeled
his own discussions of chemical experiments on the essays of Montaigne.[7]
Newton abbreviated the form to a sort of plot summary of events and findings, with himself, the "I," as narrator. And so, in some part, the scientific
article has remained. The witnessing "I" was thus science's first storyteller.
It was a way to "prove" rhetorically that the work had actually been done.

What happened thereafter, during the next three centuries, is a complex tale in itself. Different fields evolved somewhat separately, while sharing the article format and an overriding idea of what "scientific style"
should be.[8] Yet literary elegance clearly had a place in science as recently as
the end of the 19th century. Note, for example, a passage from the famous,
aether-destroying paper by Michelson and Morely published in 1887:

6.   T. Sprat, *The History of the Royal Society of London, for the improving of Natural Knowledge*, 4th ed. (London: J. Knapton, 1734), 112.

7.   These connections are discussed in J. Paradis, "Montaigne, Boyle, and the Essay of Experience," in *One Science: Essays in Science and Literature*, ed. G. Levine (Madison: University
of Wisconsin Press, 1987), 59–91.

8.   See my own "Notes for a History of Scientific Discourse," chap. 2 in *The Scientific Voice*
(New York: Guilford, 1996).

If the earth were a transparent body, it might perhaps be conceded, in view of the experiments just cited, that the inter-molecular aether was at rest in space, notwithstanding the motion of the earth in its orbit; but we have no right to extend the conclusion from these experiments to opaque bodies. . . . [And] as Lorentz aptly remarks: "Quoi qu'il en soit, on fera bien, à mon avis, de ne pas se laisser guider, dans une question aussi importante, par des considérations sur le degré de probabilitié ou de simplicité de l'une ou de l'autre hypothèse."[9]

This, indeed, seems a long way from the likes of today's article on superstring theory or quantum chromodynamics. When was the last time you read (or wrote) a paper stating "we have no right to extend our conclusion" or quoting French? What would a contemporary editor do to such a passage?

Yet there is much else that has remained in place. Don't we still propose hypotheses in order to confirm or destroy them? Don't we cite the competition, or our immediate predecessors, in a manner that supports our approach and conclusions? Of course we do, though in more formalized fashion. What, then, of the Newtonian "I" and its fate over time? Was it really killed off, forced into extinction by a more objective style? In reality, no. Both rhetorical approaches have existed side by side, and even together, down through the centuries, up to the present, though, again, in stylized form:

Anno: 1775.   I cannot, at this distance of time, recollect what it was that I had in view in making this experiment; but I know I had no expectation of the real issue of it. . . . If, however, I had not happened . . . to have had a lighted candle before me, I should probably never have made the trial.[10]

1903.   The results of the investigation of radio-active minerals . . . led M. Curie and myself to endeavour to extract a new radio-active body

9.   A. A. Michelson and E. W. Morely, "On the Relative Motion of the Earth and the Luminiferous Aether," *The London, Edinburgh, and Dublin Philosophical Magazine and Journal of Science*, 5th ser., December 1887, 450, reprinted in W. F. Magie, ed., *A Source Book in Physics* (Cambridge, MA: Harvard University Press, 1965), 369–377. The French translates as, "Whatever the case, with respect to a question of such importance, one would do well in my opinion not to be swayed by considerations regarding the degree of probability or simplicity of one or another hypothesis."

10.   J. Priestley, "Of Dephlogisticated Air, and of the Constitution of the Atmosphere," in *A Source Book in Chemistry, 1400–1900*, ed. H. M. Leicester and H. S. Klickstein (Cambridge, MA: Harvard University Press, 1952), 120.

from pitchblende. Our method of procedure could only be based on radio-activity, as we know of no other property of the hypothetical substance. The following is the method pursued for a research based on this property.[11]

1999. We first searched for neurons exhibiting a relatively high rate of spontaneous activity when the animal's eyes were closed. Next we characterized the orientation tuning properties of these neurons and selected the neurons with sharp tuning preference and robust response. We chose orientation tuning . . . because the majority of neurons in cat striate cortex are tuned for the orientation of bars or gratings.[12]

If the confessional "I" has turned into the royal scientific "we," the first-person point of view is still an important element in our efforts at persuasion. Yes, our language has tended to exchange tasteful tweed first for gray flannel and then bleached lab coat. Yes, we no longer write for someone who might be interested in an artful, novelistic type of narrative. But note how Priestley's confession of serendipity ("If, however, I had not happened . . . to have had a lighted candle . . .") changes for the Curies, who are "led" by "results" to perform their experiments, and how, in the final example, "neurons" are the principal performers within a symphony of choices conducted by the "we." The tales we tell are, by nature, still based on techniques whose goal it is to gain agreement and cooperation.

Role of Education

In fact, there are many techniques that we, as scientists, commonly use in our writing to convince our readers. I will go over some of these in a moment; but for now, a different point needs to be emphasized. It is this: changes in technical expression over time are not due, as is so often believed, to armies of editors, eager to tame an ever more rebellious cohort of thinkers. Language evolves because of a host of factors, not all of them well

11. M. Curie, "Radio-active Substances," in Leicester and Klickstein, *Source Book in Chemistry*, 522.

12. M. Tsodyks, T. Kenet, A. Grinvald, and A. Arieli, "Linking Spontaneous Activity of Single Cortical Neurons and the Underlying Functional Architecture," *Science* 286, no. 5446 (1999): 1722.

identified, and very few of them planned. As the examples above suggest, individual scientists, editors, publishers, and institutions all have played and continue to play a role. Manuals on writing and usage do, too; as noted in chapter 1, these have often been attempts to bring down the gavel and compel order, always without success. The task, in some sense, appears at once heroic and impossible. In another sense, however, it is merely part of the process.

Until quite recently, roughly the last 75 years or so, modern scientists were educated to acquire the skills of good research *and* good writing. In the 18th and 19th centuries especially, scientists were regularly among the most eloquent authors of the day (one thinks of Sir Charles Lyell or Thomas Huxley in England, for example). This was due to the type of classical education then in effect, whereby all students at the middle and upper levels studied authors of the Greco-Latin tradition, as well as grammar and rhetoric, and the works of the most successful modern thinkers and writers as well. Learning to write in a variety of styles was part of this education, and a demanding part at that.

During the past century, such training has given way to one that is far more specialist in design, far less interested in the study of rhetoric. Again, the reasons for such a change are many and cannot be explained by a few pretty thoughts or ugly phrases (e.g., "Little science has become Big science"). The background required to train in a research discipline has broadened and deepened enormously. Learning complex methods and acquiring a huge technical vocabulary (a form of language learning) are but two of many requirements not faced by our predecessors.

We can think of it another way. The amount of research performed and published today is many orders of magnitude greater than even 50 years ago (for most of the 20th century, the volume of scientific literature doubled every 15 years).[13] As this has happened, the length of the average article has tended to shrink and its style has become ever more jargon-rich: think of the papers published in *Science* or *Nature* (e.g., the last excerpt). These two journals, the most international in all of science, provide wonderful examples of what has happened to technical literature. In one way, these periodicals are holdovers from the past—they publish news and research from a wide array of disciplines. Yet in every other sense, they are leaders in language specialization. Their papers are commonly under five

13.   See D. J. de Solla Price, *Little Science, Big Science* (New York: Columbia University Press, 1963).

pages and very challenging or impossible to understand (even when well written) except to practitioners. In style, they read like ingots of terminology, and in content, they are often models of pioneering, influential work.

Why is any of this important? Because the past offers perspective and guidance. It reveals that, as scientists, we have become disconnected from the study of language, including our specific discourse, such that writing is no longer a core part of our training. Thus, the real job of a guide like this one is itself historical, that is, to help put back some of this knowledge and awareness. The past also shows that technical language is dynamic, changing. We need to write and speak in the stylistic idiom of our time, not as our scientific grandparents did generations ago. Our best teachers are our contemporaries, our colleagues (broadly defined) extending back a few decades at most. Whether we take writing classes or no, we ultimately learn to express ourselves, in our specialist contributions, from our peers, and that we can improve our communication skills through self-directed effort, instead of betting against chance and looking for figures in the carpet.

Scientific Rhetoric: An Instructive Analysis of a Notable Paper

For most of the modern era, science and literature openly employed many of the same techniques to persuade their readers. What about today? To answer this, I'd like to take a brief look at a scientific paper of some renown, published in the journal *Nature*, put it under the microscope, and point up some of the rhetoric it employs. We begin with the first few sections.

> We wish to suggest a structure for the salt of deoxyribose nucleic acid (D.N.A.). This structure has novel features which are of considerable biological interest.
>
> A structure for nucleic acid has already been proposed by Pauling and Corey. They kindly made their manuscript available to us in advance of publication. Their model consists of three intertwined chains, with the phosphates near the fibre axis, and the bases on the outside. In our opinion, this structure is unsatisfactory for two reasons: (1) We believe that the material which gives the X-ray diagrams is the salt, not the free acid. Without the acidic hydrogen atoms it is not clear what forces would hold the structure together, especially as the negatively charged phosphates near the axis will repel each other. (2) Some of the van der Waals distances appear to be too small.

Another three-chain structure has also been suggested by Fraser (in the press). In his model the phosphates are on the outside and the bases on the inside, linked together by hydrogen bonds. This structure as described is rather ill-defined, and for this reason we shall not comment on it.

We wish to put forward a radically different structure for the salt of deoxyribose nucleic acid. This structure has two helical chains each coiled round the same axis (see diagram). We have made the usual chemical assumptions, namely that each chain consists of phosphate di-ester groups joining β-D-deoxyribofuranose residues with 3',5' linkages. The two chains (but not their bases) are related by a dyad perpendicular to the fibre axis. Both chains follow right-handed helices, but owing to the dyad the sequences of the atoms in the two chains run in opposite directions. Each chain loosely resembles Furberg's model No. 1; that is, the bases are on the inside of the helix and the phosphates on the outside. . . .

The novel feature of the structure is the manner in which the two chains are held together by the purine and pyrimidine bases. The planes of the bases are perpendicular to the fibre axis. They are joined together in pairs, a single base from one chain being hydrogen-bonded to a single base from the other chain, so that the two lie side by side with identical z-co-ordinates. One of the pair must be a purine and the other a pyrimidine for bonding to occur. The hydrogen bonds are made as follows: purine position 1 to pyrimidine position 1; purine position 6 to pyrimidine position 6.

If it is assumed that the bases only occur in the structure in the most plausible tautomeric forms (that is, with the keto rather than the enol configurations) it is found that only specific pairs of bases can bond together. These pairs are: adenine (purine) with thymine (pyrimidine), and guanine (purine) with cytosine (pyrimidine).

In other words, if an adenine forms one member of a pair, on either chain, then on these assumptions the other member must be thymine; similarly for guanine and cytosine. The sequence of bases on a single chain does not appear to be restricted in any way. However, if only specific pairs of bases can be formed, it follows that if the sequence of bases on one chain is given, then the sequence on the other chain is automatically determined. . . .

The previously published X-ray data on deoxyribose nucleic acid are insufficient for a rigorous test of our structure. So far as we can tell, it is roughly compatible with the experimental data, but it must be regarded as unproved until it has been checked against more exact results. Some of these are given in the following communications. . . .

It has not escaped our notice that the specific pairing we have postulated immediately suggests a possible copying mechanism for the genetic material.[14]

One of the most epochal papers in all of 20th-century science, Watson and Crick's "A Structure for Deoxybribose Nucleic Acid" defies nearly every major rule you are likely to find in manuals on scientific writing. It does not follow the IMRAD (introduction, methods, results, and discussion) structure. It is descriptive, not analytical. It contains many "unnecessary" phrases ("in other words," "so far as we can tell,"), vague words ("ill-defined," "loosely"), and redundancies (the first sentence is repeated at the beginning of the fourth paragraph). There are even grammatical errors (frequent use of the unrestricted "which" instead of "that"). Its paragraph form is uneven and improper. It has not included or even consulted the relevant experimental data (and admits this!). And finally—but this is far from a complete list—it has no real conclusion.

Yet it works. We can agree that, despite all its unforgivable blunders, the article is persuasive and effective. Why? First of all, it answers quite well the five fundamental questions essential to good scientific storytelling:

1. *What* did you (and your coauthors) do?
2. *Why* did you do it?
3. *How* did you do it?
4. What did you *find*?
5. What does it *mean*?

For question 1, Watson and Crick say they worked on the structure of DNA. They then tell us why (question 2): it has "considerable biological interest" and hasn't been solved by anyone else, other models being "unsatisfactory." Their answer to question 3 is less clear, and we have to deduce it through the logic they provide regarding the details of their structure, as well as the strong suggestion that they used X-ray data from crystallized DNA fibers. Their paper, however, was published in 1953, a fact that accounts for some of the stylistic differences from today's discourse (an inter-

---

14.   J. D. Watson and F. H. C. Crick, "A Structure for Deoxyribose Nucleic Acid," *Nature* 171, no. 4356 (1952): 737–738.

esting exercise: rewrite the paper so that it sounds more contemporary to the 21st century). Today, reviewers would likely demand, with reason, that the authors be more explicit about their data and methods. As for question 4, the answer is the model itself, which is described and also shown in the paper's single figure, that of the double helix—an image, we might note, whose simple elegance has helped it become nothing short of an icon for the entire field of genetics. Finally, Watson and Crick tell us the deeper meaning (question 5) of the structure in the last sentence, rather coyly, we might note, but distinctly.

There is another way to look at all of this. We can consider what the article actually *does*, step-by-step. It has a flow that moves through the following stages: (1) a declaration of its importance and novelty; (2) a statement that its problem has not been solved, even by some worthy competitors; (3) discussion of its findings (paragraphs 4–7); (4) qualification of these findings, as not yet complete or final (need for further work); and (5) specification and elevation of the importance declared at the beginning. What type of "story" does this offer? A tale full of understated sound and fury. The authors tell us that there is a rush to unravel the "grail" of DNA, that all important alternatives appear doomed, and that they alone have happened upon the rightful path. Watson and Crick then reveal their discovery, build a castle out of detailed descriptions, and set their flags flying. A final fanfare announces that their victory holds yet a greater triumph than first imagined. Like a well-crafted story, the argument follows an hourglass structure overall: it begins with the general, moves into the ever more specific, and then, at the end, expands back outward into the general.

Now, let us look a little closer. Note that the paper begins with short, concise sentences, which evolve into longer ones as the authors begin to describe their model. This progression, in fact, matches the main effort of the paper, which is to "build" the model in words, through specific details. The use of brevity at the beginning moves the story along more quickly, and us with it. Shorter sentences at the end provide a touch of subtle drama and a more memorable finale, so to speak. We, as readers, are therefore treated as if we deserve to see only the best, most privileged information; we are given the authors' full confidence.

Notice, too, the use of "we." This also happens at the beginning and the end. Why not in the middle? Because this is where the main subject of the narrative becomes the model and its structure. We can envision this as a video that starts out with the authors in frame, then pans to the model, moves in for a close-up, pulls out, then shifts back to the authors. In the article itself, this happens so smoothly that we don't perceive it, though it is a true rhetorical technique, such "invisibility" being the sign of a good paper.

Finally, the paper employs a form of logic that is speculative, yet, at the same time, as ordinary in scientific communication as the comma. We see it in the form "if we assume . . . we then find . . ." (paragraphs 6 and 7). Speculation, say many writing guides, has no place in technical writing except perhaps in certain arm-waving theoretical communications. But they are wrong. The posing of scenarios, possibilities, and alternatives is a venerable and valuable type of reasoning in science. The "if . . . then" technique acts like a postulate and is part of the rhetorical toolbox of the competent scientific writer, past and present. We see with Watson and Crick that it is used to propose what they think is actually the case, but for which they lacked the data to firmly claim in a publication.

At this point, you might feel we've gone far enough with our analysis of this brief, two-page paper—indeed, maybe a little too far. We've made Watson and Crick's article appear no less complex than a protein molecule or fusion reactor. Yet the truth is that a full rhetorical study of "A Structure for Deoxybribose Nucleic Acid" would easily fill an entire monograph. Wouldn't such an analysis be wasteful for our purposes? No, not really; just unnecessary. From the above, it's obvious that a great deal is going on in the average scientific paper. The surface may seem calm and composed, like a windless lake, yet only a few drops beneath the microscope are enough to reveal a world of strategy, claims-making, and enlistment.

This begs a question: are good writers aware of the techniques they employ? The answer is almost always yes and no. Yes, because good writers very often plot out or experiment with the logical course of their narrative. No, because many specific rhetorical techniques are used intuitively; they have been learned and internalized by attentive reading of the literature. Effective writers are those who have an inner ear for what sounds right, what is persuasive at each turn of a discussion. Being aware of even a few such techniques, and how to acquire them, will provide the scientist with a powerful instrument for his or her expressive work.

Grammar: Facts and Fallacies

Questions of history and language change bring us, happily or unhappily, to the subject of grammar. What is to be said about it? Look again at the first set of examples given in this chapter: notice that, as the number of technical terms has grown, the grammar of the sentences has become more simplified. This is not an accident. Consider the following:

> A comprehensive overview of quality control in DNA would include a discussion of DNA polymerase fidelity and postreplicative mismatch correction and would also consider the damage-responsive cell-cycle checkpoints and the signal transduction systems that lead to cellular effects.[15]

Now replace each technical term or phrase with an ellipsis:

> A comprehensive overview of ... would include a discussion of ... and ... and would also consider the ... and the ... that lead to cellular effects.

Finally, with a bit more distillation, we get

> A comprehensive overview of ... would include a discussion of ... and would also consider ...

Once the terminological smoke clears, the average scientific sentence today emerges as fairly elementary. Perform the same exercise on an entire paragraph in a recent article—on your own writing even. I guarantee it will be revealing, and helpful. Indeed, the exercise can be valuable in pointing out unneeded complexity (the above sentence, for example, could certainly be helped by enumeration).

I ask that you neither let the subject of grammar cripple you with concern nor float your intentions too high. Good grammar alone does not a writer make; bad grammar, if only occasional, does not destroy one. On the other hand, to communicate effectively, you have to be able to produce a legible sentence. This goes (almost) without saying. But being hyperconscious of possible mistakes can reduce your progress to a glacial melt. Perfectionism makes for unforgiving standards, whether you are a writer or reviewer.

Grammar is mainly about rules and formulae. It necessarily involves a rule-driven, mechanistic view of language use. It is the zero law of the communication process: necessary but way far of sufficient. "Proper usage," on the other hand, is, in reality, a different subject. Authors of technical

---

15. T. Lindahl and R. D. Wood, "Quality Control by DNA Repair," *Science* 286, no. 5446 (1999): 1898.

writing manuals are often in riot gear over whether to change "prior to" to "before," "perform" to "do," and so forth. Such efforts are not about grammar; they are about brutal simplification as a standard. They are largely ineffectual and, worse, irrelevant. In many cases, they represent pet peeves of individuals. They are important only when they belong to journal editors and, to a lesser degree, reviewers. In general, scientific discourse rumbles on, fortissimo, more productive than ever, without paying much attention to these constabulary proclamations.

Does this mean "standards" are out the window? Is science groaning under the weight of its own fatty, acidic verbiage? Hardly. Yes, there is poor writing in abundance (as in other fields); and indeed, it is important to do something about it. But implying that a few dozen ironclad rules will improve scientific expression overall is tantamount to believing that knowledge of the periodic table automatically makes you a chemist.

Thus, my advice is this. If you feel shaky about your ability to generate a correct sentence, to the point where it slows or even blocks your writing, then by all means focus some initial effort in this arena. Study a grammar text for a month or two, or more. Do this while reading the literature of your field with an occasional eye to evaluate its grammatical aspects. Gain a degree of confidence here; lessen your concerns. Learn the basics before proceeding on to the delights, challenges, and deeper humiliations of professional writing.

What It Means to Write or Speak Well in Science

When it comes to putting words on paper, the scientist has a certain advantage. What, possibly, could this be (don't all scientists *hate* to write)? Simply this: at a basic level, scientists have a choice generally denied to other disciplines. They can be purely functional writers—ordinary engineers of the word—or they can strive for higher levels of eloquence, even creativity, within the bounds of acceptable formality.

Outside of science, functional writing is commonly looked upon as dull and unskillful. In most fields, there are expectations (or hopes) of grace, color, style. Particularly in the arts, in history or literature, language is expected to call attention to itself, to at least strive now and then for some obvious sign of craft, cleverness, or felicitous phrasing. Even in business or sociology, material with the aesthetic quality of cement is viewed as lacking in something. Not so in science. We can be as flat and gray as we like and not be judged ill for it. Functional writers, in fact, compose the great majority of successfully published scientists. This does not mean that such writing is always good—but a significant portion *is* competent: readable, informative, and adequately organized. Moreover, few writers are bad (or good) all the time; varieties of competence tend to exist in any single piece of writing. Functional communication, at a proficient level, is very much something to strive for in science. Indeed, as we've already indicated, it is essential for good science to be done.

There is something else here, too. Nonfiction authors frequently come up against a number of fundamental questions. William Zinsser, for one, has laid these out nicely, showing how they include (but are not limited to) the likes of, How am I going to address the reader? (Reporter? Provider of information? Average man or woman?) What pronoun and tense am I going to use? What style? (Impersonal reportorial? Personal but formal? Personal and casual?) What attitude am I going to take toward the material? (Involved? Detached? Judgmental? Ironic? Amused?) . . . Who will my readers be? What sorts of publication venues might be interested in my work? How much will I get paid?

Nearly all such questions are irrelevant in science. We simply don't need to worry about them. They've already been answered, in large part, by history. Such is a benefit to professional discourse in almost any field, but much more so in science. Again, this does *not* mean that our speech is simple and unchallenging, not in the least. We have our own problems to solve that other nonfiction writers lack: how to translate data into words; how to describe experiments so that they might be repeated; how to use illustrations; which colleagues to cite or challenge.

But can we really choose to be eloquent writers in science? Assuredly we can. Here is where aspects such as refined organization, use of transitions, sentence rhythm and length, and strategic employment of rhetorical technique come in. An entire chapter of this book is devoted to showing how some of these aspects can be used creatively. The trouble is that such creativity needs to be subtle in science. It reveals itself most often in an occasional manner and in background elements that direct and propel the argument, but quietly. It may therefore go largely unrecognized by a majority of readers. This is the risk you run in crafting beautiful science: only editors, writers such as yourself, and teachers of writing are likely to appreciate what you've done, at least at first. In the long run, of course, you are also in jeopardy of being used as a model for others.

I began this chapter with the contemporary scientific author, and it makes sense to end this way. History proves that, as writers and speakers, we are immediate contributors to the evolution of scientific discourse. Every article or proposal or report that we produce, every word we put to paper, is an event within the flow of this evolution. Scientific language continues to change, as it must. We are its creators and metabolizers. But even more, we are its primary medium: through our efforts to create and exchange knowledge, this language is made real and alive. To write and speak well, whether functionally or eloquently, is to take responsibility for history, for knowledge, for oneself as a scholar.

3. READING WELL: THE FIRST STEP TO WRITING WELL

The only demand I make of my reader is that he should devote his whole life to reading my works. —JAMES JOYCE

The Concept of Authorial Ear

Good musicians and skilled writers have something in common. They both have developed an ear for what sounds right and what does not. When faced with a work of music or text, the well-tuned ear can listen for the movement of notes and words, certainly, but it will be equally alert to patterns of sense, to elements of order and logic and how they move within the work. Having this type of skill is no small thing. But it is no mystery either. It involves being attentive to the medium in particular ways. It means, for a scientific author, being able to detect what feels awkward in a sentence like this:

In higher plants, flowering—the transition from vegetative to reproductive growth phase—is controlled via several interacting pathways influenced by both endogenous factors and environmental conditions.

This is the opening to a published paper in a prominent journal. It is not a wonderful sentence, but it is comprehensible. Suppose, however, it were written like this:

Flowering in higher plants is defined as the transition from vegetative to reproductive growth phase. This change is controlled by several interact-

ing pathways, each of which is influenced both by endogenous factors
and environmental conditions.

Breaking the original sentence up like this creates a very different sound
and flow. The information becomes more pleasant to read, easier to remember. A short introductory definition of a fundamental process is followed by a longer sentence, explaining its controls. Sound and sense go
together; one mirrors the other. Ideally, the next sentence would be a bit
longer and would either explain the noted pathways or introduce a species
of plant whose growth activity has been investigated within the context
just presented.

Every piece of music and each scientific article is an attempt to transfer
something to the audience, not merely to "express" or "publish" it. Writing,
in particular, is always a form of teaching—an attempt to give the readers
what they did not have before. Being sensitive to this process counts as an
invaluable advantage. Recognizing good writing for what it is can be the
first step to actually doing it yourself.

Internalizing Preferences: The Value of Models

Is it possible for the "ordinary" scientist to acquire such an ear? Musicians,
we know, have talent. But talent must be trained and developed and, in
any case, can be partly defined as a heightened ability to imitate and go
beyond provided material. What about writers? Certainly talent, though
vaguely understood, may exist here too. But skilled writers in a great majority of cases—and especially in the professions—are made, not born.
How? The most natural and effective process, and the one most often followed throughout history, is that of apprenticeship. Such, indeed, is really
how any complex skill is acquired, sooner or later.

We first learn how to do research this way, by imitating teachers, older
colleagues, and so on. Throughout our career, moreover, we are likely
to absorb tips and techniques from our betters and those that help us in
our work. Apprenticeship involves disciplined emulation; you gain ability
over time by incorporating and adapting the good work of others, so that
you eventually develop a style of your own. Even the most gifted musician
moves through a succession of composers and pieces, writing composi-
tions that are at first wholly imitative, often painfully so. A personal style

emerges as the composer or writer internalizes and mutates what she or he has selected as most worthy or worthwhile.

A deep difference between science and music is that, for scientists, this type of learning is ordinarily scattered, sporadic, unsystematic. Researchers are often left to do it themselves, despite the benefit of one or two writing courses during their college career. Imagine if this were true for musicians, journalists, or historians, however. Yet the scientist, whose work is no less involved in putting ideas down on paper, is largely abandoned to his or her own devices here.

That being said, where does one begin? Lacking a mentor or journeyman on hand, we turn to their embodiment—examples of solid, successful writing. Apprenticeship here involves two main activities: first, collecting examples of especially good writing whenever you come across them, and second, going over them in an attentive manner, so that, sooner or later, they become internal guides for your own sense of what sounds good and what doesn't.

The main thrust is to identify models of effective writing, study them, and then find ways to emulate them. In choosing your examples, you might select entire articles or only parts thereof, whether individual sections, paragraphs, or illustrations. In studying what you have chosen, you might begin by trying to understand what it is that appeals to you: Is it the flow and rhythm of the sentences, clarity of the wording, an inventive argument, the visual organization of an illustration, nicely turned phrasing, all of the above? To emulate, meanwhile, you can reread your models, copy them out, memorize and recite them, write a paragraph in their style. Many techniques exist. But the final goal is the same: make these models your own, internalize them. This is how you can put them to work for you.

One final point. Manuals on technical writing, and editors too, have often emphasized how much bad writing there is in science. "All are agreed," said a one-time editor of *Science*, "that the articles in our journals—even the journals with the highest standards—are, by and large, poorly written."[1] Actually, I am one who does not agree (and I know many who also do not). This statement is a drastic oversimplification and, in fact, something of an insult to the good writers who do exist. No need to justify a moral high ground for those who, in the midst of their fatalism, would have

---

1. F. P. Woodford, "Sounder Thinking through Clearer Writing," *Science* 156, no. 3776 (1967): 744.

us all be scientific Shakespeares. Really good writing, admittedly, is rare in most fields of endeavor—but this doesn't mean it is absent, not in the least. Seek and ye shall find it.

Being a Critical Reader

To identify models, we need to read others' work in a particular way. We need to be alert to rhythm and sense, as well as logic—all in all, how the "story" is told. To do this may take a bit of practice. Scientists are taught to look for "content" above all else; this is how we have been socialized to read. But every piece of writing has multiple levels of sense within it, as our examples in chapter 2 revealed. When we read a beginning like this,

> Galactic dark matter may consist of weakly interacting particles which can be captured and trapped in stars, and which would then contribute to the transfer of energy. A special class of these particles ("cosmions"), with weak cross-sections that are larger than standard has been invoked as a solution of the solar-neutrino problem, and also as a means of suppressing convection in the cores of horizontal-branch stars.[2]

we know, if only at the back of our minds, that we are in for a bit of heavy weather. Too much is being packed into these sentences; the reader is being treated as a highly absorbent material. We find ourselves having to reread in order to get what we need. Now look at another article opening:

> Subduction of the Juan de Fuca and Gorda plates has presented earth scientists with a dilemma. Despite compelling evidence of active plate convergence, subduction on the Cascadia zone has often been viewed as a relatively benign tectonic process. There is no deep oceanic trench off the coast; there is no extensive Benioff-Wadati seismicity zone; and most puzzling of all, there have not been any historic low-angle thrust earthquakes between the continental and subducted plates.[3]

2. D. Dearborn, G. Raffelt, P. Salati, J. Silk, and A. Bouquet, "Dark Matter and the Age of Globular Clusters," *Nature* 343, no. 6256 (1990): 347.
3. T. H. Heaton and S. H. Hartzell, "Source Characteristics of Hypothetical Subduction Earthquakes in the Northwestern United States," *Bulletin of the Seismological Society of America* 76, no. 3 (1986): 675.

A passage like this makes it clear that technical language can be a pleasure. The writing has pacing, flow, and elegance. There is a diversity in vocabulary ("dilemma," "compelling," "benign," "puzzling") that draws us in, emotionally and intellectually. The sentences set up an excellent rhythm (short, medium, and long with breaks) that moves us along, deeper into the material. And beyond all of this, the words simply sound good to the ear. Everything about the writing makes us want to read further, with anticipation.

These are the sorts of aspects that we need to be sensitive to, as critical readers. There is more than this, however. If an article or paragraph seems well-written, like the above example, take a moment or two to look at how it is organized. In the last excerpt, for example, notice how the writing moves from the general to the specific, how it adds detail to the topic at issue. Notice too that what is being discussed is the *absence* of certain phenomena—an excellent way to establish the "problem."

This type of analysis is good for short sections that you find appealing. For whole articles that seem well-written, on the other hand, you might check the overall logic by reading through the headings, surveying the illustrations, looking at the conclusions, getting an idea of how well it all fits together. Very often, you will find a mixture of things: admirable passages separated by mediocre writing; good style compromised by hurried organization; poor text offset by high-quality illustrations. More rarely, you will discover an article, report, or proposal in which all elements work together in excellent fashion. Save it, study it, make it part of your repertoire.

A Few Issues to Consider

Scientific articles and reports are commonly made up of different sections—abstract, introduction, methods, and so forth—many of which differ among fields. Each of these sections takes up a distinct type of content and, to a significant degree, uses a distinct style of writing, a different voice. A good introduction sounds different from a methods section, which in turn should be stylistically separate from a discussion of results. An article tells its story by assembling these different voices and weaving them together. Transitions are essential (look for them!), but seams remain. Recognizing such complexity is part of the work you should do in perusing the literature. It shouldn't dismay or overwhelm you in the least—after all, it is achieved every day by thousands of scientists, in every field.

Students and unpublished scientists may feel that nearly all material in the major journals is worthy. After all, it got published, didn't it? But articles appear in journals for a number of reasons, and quality of writing (unfortunately enough) is often not among them. A fair (or unfair) amount of bad text gets a day in the sun. Editors have demanding jobs and cannot possibly rewrite every piece of poor writing on important topics. Thus, a large amount of flawed text finds its way into even the best journals. You might check the howlers provided by Vernon Booth in *Communicating in Science* (1993) and Robert A. Day in *Scientific English* (1995). As they note, the most common slipups are the improper adjective, for example, "mountain-building geoscientists" or "infant experts" (i.e., researchers with grand ambitions vs. babies with brains), and the infamous dangling modifier: "Having been placed in an oven at 575°, we dried the sample for 3 hours" (no comment, in memoriam). I would suggest that you collect some of these, too, as negative models, useful reminders. Nothing succeeds like the threat of humiliation.

Critical reading therefore means making judgments, being judgmental. We are trained as scientists to be evaluative, even skeptical, of each other's work. We need to extend this into the realm of expression, too. Perhaps the most simple, and effective, way to initially judge a piece of writing is to ask yourself: is this something I wish I had done myself, or am I glad I didn't? Once you have answered this question, you can go on to analyze the reasons for your response.

Some Techniques to Consider

The following are some suggested methods for choosing and using models. Note that I am *not* advising you to use all these methods. Different techniques will be useful (or possible) for different people and situations. Use what follows according to your needs, available time, and inclination. You should feel free to modify any of them or even come up with your own.

Choosing Models

1. To gain a better feeling for the sound and rhythm of language, read high-quality older literature in your field, say from 100 years ago (e.g., Pasteur in biology, C. G. Gilbert in geology, Kelvin in physics, Gibbs in chem-

istry, etc.), and compare this with a contemporary article or book. This exercise will help tune your ear to differences in language flow. (Note: this is not intended to suggest older literature as a model of style for you, only as a type of counterpoint to help sensitize your ear).

2. If you're unsure about how selective to be, begin by making a fair number of choices and then whittle the totality down, bit by bit, as you reread, to the very best. Over time, you'll find that certain selections will continue to impress, whereas others will lose their luster. If there are authors in your field who are known to be good writers, you might check their work first. Don't, however, confuse abundant publication with good authorship. You are looking for examples of superior expression, not impressive credentials. Whatever strikes you as particularly well done is worth collecting.

3. Note the authors of examples you have chosen; read other work by them to see if it is of equal quality. You may find that you are especially attracted to the writing of only a few authors. This is fine; it will help make your selection easier.

4. Think about forming a reading group or reading seminar with some of your colleagues. A brief (e.g., one-hour) session each week to discuss the current literature in your field would be a good basis for evaluating and selecting examples of good (and bad) writing. This is something that is done fairly often in the humanities but, in my experience at least, very rarely in the sciences.

5. Make sure that you know what venue your samples came from. Textbook writing is not the same as journal writing, which is not the same as proposal writing. Each has its own specific requirements for writing style and logic. Be alert to such differences.

6. If possible, eventually choose a small number of models (5–10) that you find particularly admirable. Stick with these for a while, until your judgment matures. Then think about replacing some of them with newer examples when they come along. Most important of all is to choose examples that you wish you had written

Using Models

1. Reread your chosen models on a regular, or fairly regular, basis. Make it a habit to go over them, if only briefly, at particular times when you have a few minutes free.

2.   Try reciting out loud or copying in longhand those passages that you especially admire. This is a time-honored technique that helps sharpen your awareness and makes each passage more immediate.

3.   Choose one sample, say the first paragraph of an article introduction, read it through carefully several times, paying close attention to such things as sentence length, word choice, and use of punctuation. Write a following paragraph imitating the same style (it doesn't matter whether the information is accurate or even real). If this is difficult, then copy the model paragraph directly, changing some of the words.

4.   Try taking a document of your own and rewriting a paragraph or two in the style of your chosen sample (i.e., as you might expect your selected author to have done).

Writer's Block: A Different Perspective

Writer's block is a subject on which innumerable authors have weighed in. Ideas about its nature and origin are legion; solutions, however, go wanting. In part, this is because writing—of any type—is a very personal affair, thus so are inhibitions and anxieties associated with it.

Here I define this state as one in which all forward progress in getting words on paper is halted. This might happen when one first sits down to write, when there is a struggle to find the right word, when earlier composed material seems hopelessly bad, or when the entire effort feels doomed. Different authors have developed different methods for dealing with such interruptions. Some methods are mechanical (e.g., "Leave a space for the missing word and go on," "Skip several lines and start on the next idea," etc.); others are ecological ("Clean and reorganize your office, take a walk in the country, go for a run, get married, then sit down again"). One well-regarded source dodges with the problem in this way: "There are as many kinds of writer's block as there are writers . . . and my name isn't Sigmund Freud."[4]

I would like to propose something a little different. Writer's block, we might say, occurs when the authorial ear—that inner voice of ours that brings forth words—is stopped or jammed. Something is needed to help

---

4.   W. Zinsser, *On Writing Well: The Classic Guide to Writing Nonfiction*, 6th ed. (New York: Harper Reference, 1985), 23.

that voice begin speaking again. Let me suggest a few methods for doing so, along the lines of what I have been saying:

- *Go back to your models* and read through selected passages on topics as similar to your own as you can find. Recite or copy them out, if necessary.
- *Read a past article of your own,* preferably on a subject not too distant from the one you're working on.
- *Discuss your article or research with a colleague.* Use this as an opportunity to explain the particular point or section where you're stuck. Listen for useful phrases in your descriptions and explanations; jot them down.
- If writer's block occurs when your article is already partly written, *read what you have thus far,* from the beginning up to the ending point. This may help gain new momentum.

The basic idea is to find a way to restart the flow of language and confidence within the authorial ear. None of the above can be considered fail-safe. The notion of reciting out loud or copying in longhand someone else's work might seem childish or embarrassing; if so, don't do it, except perhaps as a final resort. A fair number of my students have found that simply reading through articles on similar topics helps defrost the fluid of language. Others prefer to get closer to the actual words by hearing themselves say them or write them out. In extreme cases, I have advised an author to find a paragraph or section very close to the one he or she is stuck on, in subject matter and intent, and to rewrite it, sentence by sentence, changing phrases and terms where necessary but leaving a fair bit of the original intact. This is where "the common language of science" (as Einstein called it) can come to the rescue.

T. S. Eliot and the Importance of Theft

The famous poet T. S. Eliot said that immature poets borrow; mature poets steal. This statement is well-known among professional writers but very few wish to admit its truth directly. Using others' work as a template for one's own, however, is as old as authorship itself. It is not actually stealing, of course (unless you do plagiarize), but rather a form of learning from others. Adopting tricks of the trade from colleagues is done everywhere else in research—why not in writing?

If you see an organizational scheme, a paragraph structure, a phrase that you especially like, make it your own. You might, for example, find the type of introduction given in the passage above on subduction of the Juan de Fuca and Gorda plates particularly effective (I do). If so, imitate it directly; add it to your toolbox. Here's an example to show what I mean:

> Model passage.   Subduction of the Juan de Fuca and Gorda plates has presented earth scientists with a dilemma. Despite compelling evidence of active plate convergence, subduction on the Cascadia zone has often been viewed as a relatively benign tectonic process. There is no deep oceanic trench off the coast; there is no extensive Benioff-Wadati seismicity zone; and most puzzling of all, there have not been any historic low-angle thrust earthquakes between the continental and subducted plates.

> Adoption.   The precise mechanisms by which granite is emplaced in continental crust have continued to elude researchers. Despite abundant evidence for intrusion along fault segments, emplacement has been typically viewed as a relatively passive or nontectonic process. Granites are commonly undeformed; they show few signs of internal lineation; and, above all, their contacts with surrounding rocks usually exhibit only small amounts of shearing.

Clearly, this is not plagiarism. It is the sort of thing that writers speak of when they mention they have been "influenced" by another author. Frankly, once you begin to read the literature with this kind of eye, looking for potential influences, you will see that *all* writers, even the most influential, constantly "borrow" and "steal" from each other. It is, in fact, the most natural thing in the world and is practiced in every form of human expression, including music and art. Much of it, to be sure, occurs unconsciously. Why not make it conscious, therefore less haphazard?

No writer or musician or artist is a Galápagos. As scientists, especially, we inhabit a mainland of both private and communal expression. Imitation, leading to adaptation, is the sincerest form of advancing our survival in this setting.

*

4. WRITING WELL: A FEW BASICS

Reading maketh a full man, conference a ready man,
and writing an exact man. —FRANCIS BACON

Functional Expression

Scientists have two basic choices as communicators. They can aim to be proficient and functional, or they can strive for higher levels of literary skill, even mastery. The first of these choices is for everyone; the second is not. Room exists in science for both, and both certainly exist.

Functional communicators are able to write and speak accurately, with reasonable precision, in a clearly organized fashion, without too many significant grammatical and syntactic errors. This mode is preferred by a majority of professionals. If achieved, it is not only acceptable, but wholly admirable. Functional communication embodies the philosophy that writing and speaking are methods for making knowledge available in an efficient, usable manner. This philosophy is a good one—as long as it stays within its own, limited context (as a general outlook, however, it is likely to be stultifying).

To write proficiently, you need a number of skills: a sense of good and bad grammar, an ability to impose order (sometimes out of chaos), the power to think visually (for illustrations), an ear for monologue. You need something else too, however.

Writing demands, absolutely requires, a type of intense concentration, like chess or playing an instrument. This, too, is a skill. Scientists already possess it, in trained form, with regard to performing research or field-work. The same type of consistent focus you devote to observing and re-cording must be transferred to writing. Generating new knowledge is a creative act, with two scenes—investigation and composition.

To Write Is to Experiment

Writing is a process of experimentation. This is a crucial realism for scientists, and indeed for all professionals. Producing good, functional documents involves trying things out, engaging in trial and error, tinkering around. Even if you have a very firm and clear idea of the text you want to write, what finally emerges will rarely accord with this image in any precise way. There are simply too many elements that need to be worked out, too many levels of detail and decision making. And each of us is too much the human being, at once faulted and engaged, to act like a machine in this capacity. All of which helps us understand why writing is such hard work (for it is).

It can be an enormous help to know from the very beginning that you'll be entering this trial-and-error process. Experienced writers anticipate that there will be dead ends, struggles, and triumphs in lowercase. They know that they may even stumble across new, unforeseen ideas, those that illuminate their current project in a different way or that open a prospect for future work. Writing often involves such discovery—this, too, is part of its experimental nature.

Experimentation may continue up to the time that your document leaves your hands and goes out for review or publication. At this point, it is no longer yours: it belongs to the world. The document will now speak for itself. There is no "what I wanted to say" or "they'll know what I mean." There are only the words on the page as you left them.

I ask, finally, that you avoid one error of belief that is monstrously prevalent. This is the widespread notion that "to write clearly, you must first think clearly." This sharp little maxim may appear logical, but it is really rubbish. No matter how rational your thought may be (or appear to be) on a particular problem, no matter how detailed your intentions and plottings, the act of writing will almost always prove rebellious, full of unforeseen difficulties, sidetracks, blind alleys, revelations. Good, clear writing—writing that teaches and informs without confusion—emerges from a process of struggle, or if you prefer, litigation. This is true irrespective of how experienced an author may be, how many dozens or hundreds of papers he or she may have published. Most often, the terms of the formula given above need to be reversed: "clear thinking can emerge from clear writing." Imposing order by organizing and expressing ideas has great power to clarify. In many cases, writing is the process through which scientists come to understand the real form and implications of their work. But you must always be prepared for the process to be messy, experimental. If it were not, the best advice any guide could offer would be to see a therapist.

The Reader

Every piece of writing has an intended reader. This reader in science is likely to be a colleague, an interested peer who hasn't yet learned of your work. If your writing is effective, this reader will be fairly consistent.

The first few lines of a document establish the reader. An article that begins

> The nicotine acetylcholine receptor is a ligand-gated channel that mediates signaling at the vertebrate neuromuscular junction.

immediately weeds out anyone unfamiliar with the terminology given. The reader here is obviously a specialist. Starting right in with the detailed subject matter like this, though sometimes condemned, is fully accepted practice in many areas of science. Certainly, it makes sense from the point of view of efficiency.

On the other hand, a beginning like this

> In recent years, there has been a marked resurgence of interest in artificial neural networks.

or this

> We report here results from the first stellar occultation by Saturn's giant moon, Titan, ever observed.

seems written for a more general reader. In each case, the second line might be crucial:

> Such networks consist of computer systems designed to imitate, in streamlined form, certain basic principles of operation of the human brain.

Data obtained include the occultation chord at each station listed in Table 1, using half-intensity times, t, when the fraction of unocculted stellar light hitting the Earth was 0.5.

If in the first example the readers remain general, in the second they are suddenly whittled down to professional and high-level amateur astronomers.

Both of these approaches, too, are common practice. If the first is commendable for consistency, the second begins with a bit of flourish to stimulate interest, and then gets down to work right away.

The important thing is to install your true reader in the first several lines and then stay the course. Many writers make the mistake of dragging out their general openings, and then suddenly shifting to specialist discourse, or jumping back and forth between the general and specialist reader. This can frustrate and confuse the audience.

Check some of your models; see how they begin. Look at the first one or two paragraphs and see who the ideal readers are, how they are set up. Examine different journals in your field; get an idea of what is common, what works, and what doesn't. Reading for "the reader in the text" offers a valuable tool to evaluate your own and others' work.

The Author

If every document contains a reader, it also speaks to this reader with a particular voice. In science, this is necessarily a voice of authority, a giver of knowledge. Let's look again at one of the examples above:

> In recent years, there has been a marked resurgence of interest in artificial neural networks. Such networks consist of computer systems designed to imitate, in streamlined form, certain basic principles of operation of the human brain. We report here on advances in one area of current research—optical signal processing using holography.

What kind of voice is this? Hesitant? Enthralled? Confident? Clearly, the last choice applies. The author projects a tone of self-assurance; he is in command of the material. He speaks as someone more knowledgeable, someone who wishes to teach, inform, guide. This is the persona of competence in science.

Of course, there are moments in almost every text where you need to qualify what is being said, to back off a little. This is frequently the case, for example, when you are making new generalizations, discussing the implications of your work, proposing new ideas, or simply admitting the limitations of what you've achieved:

These results

suggest the possibility that previous interpretations are erroneous . . .

can be interpreted to indicate . . .

generally support the concept of . . .

are preliminary and indicate a need for further work in the areas of . . .

Knowing when to qualify what you are saying is part of being a confident author, projecting your expertise.

On the other hand, one of the most frequent complaints I have heard from editors is that beginning scientific authors are often far too hesitant in their writing. They are much more likely to say

Our work here, though preliminary in part, can be considered to support the conclusion that . . .

rather than

Our work supports the conclusion that . . .

The second example provides whatever error bars you may need. Scientists recognize such clues when they see them. Note that what the first example really says is, "Who are we to pretend that we can contribute something meaningful?" Or, as once put by a well-regarded investigator of the human condition (Groucho Marx), "Why would I ever want to belong to a club that would have someone like me as a member?"

Once again, look through your models. See how they set up their voice of authority. Pay close attention to places where facts or ideas are stated firmly and where they are softened by qualifiers. This, too, will help tune your ear and guide your hand.

Organization

The skeleton of every document lies in its organization, the ordering of its parts and substance. Good organization involves several levels of order. First, there is the sequence of major sections, most commonly some version of abstract, introduction, background (e.g., geographic setting, materials and methods, previous studies), main discussion, conclusion, references. Second, there is the order of any subsections under these major headings. Together, the heads and subheads of a text should provide a kind of table of contents—indeed, it is sometimes a good exercise to extract them and see how well things are put together. Third, the writer must work out the degree of detail to be included in each section, as well as the progression of this detail. All of which sounds like a tall order, and it is.

Many writing guides provide one or another system for helping you organize your document. The basic idea is that structure determines flow of the argument, and therefore its persuasive quality. If this were true, however, we would be able to simply dump information onto the page to tell our story. Organization can no more guarantee good composition than bones can bring a body to life. If we are to be physicians of the word, we require an equal or greater amount of biology.

No single group of methods, however carefully explained, can possibly encompass the range of needs, problems, and styles that scientific writers find individually relevant. There is no "right way" to impose order on the chaotic heap of information you've generated. Some scientists I know write best using a sketchy outline. Others require a veritable train schedule, laying out the arrival of one point after another. Still others plot one section at a time, as they come to it. Some writers, however, use no outline at all. One researcher of my acquaintance (a very skilled writer) works by pinning note cards randomly to the wall and then visually placing them into a dendritic (rootlike) structure. Different approaches succeed for different authors.

For inexperienced writers, there are a few practical techniques that might help in the early stages of composition. Making a provisional outline that includes a title for each major heading and subheading is one way to start, as long as you leave open the possibility that things may well (and often should) change as you write. If your research is laboratory-based, you might consider at least beginning with the standard structure—introduction, materials and methods, results, discussion, conclusions—and add subheads where you think they might go. You might choose only your major headings, place these on separate pieces of paper, and then write down ideas, data types, or the principle points you feel should be covered under each. If you prefer to plan things out visually, use a process similar to my friend the note-card user: write main ideas and data areas on separate cards and see what type of logical order they fall into. Another possibility is to assemble the illustrations (in rough form) you might consider using and search for a logical sequence among them that might guide the writing.

These are just a few suggestions. You should feel free to play with them in any manner you choose, ignore them altogether, come up with your own. In nearly all cases, it will take time and some experience for you to determine a style of organizing material that works well for you. It helps to be patient and to investigate a bit in this realm, too.

As an overall guide, think of your document as moving from the general to the more specific and then back to the general again—a kind of hourglass or vaselike shape. Begin broadly, that is, introducing your reader to the "problem" and its background, then discuss your methods, present your detailed findings, and finally venture outward again in your conclusions. The greatest concentration of details should be in the article's center, the discussion of what you found, what new data you generated or interpreted, what it means. In truly eloquent science, this type of hourglass approach tends to be repeated within each individual section of a document: the reader is repeatedly given a focus that works like a series of musical variations: general (allegro), specific (largo), and general again (finale: adagio). Functional communicators do not have to go this far. However, using the general-specific-general pattern to help organize your text is very effective.

Please be warned, however: there is usually no way that you can decide all levels of order in your document beforehand. Most often, it helps to begin with a general plan, whether written down or not, and then proceed to write. In a few cases, your outline may hold true for significant stretches; more often, however, it will need to be altered. Writing, as I've said, is a process of experimentation and discovery.

Above all, consult your models. Look at articles on similar topics, preferably in journals to which you might submit your work. Adopt or adapt a structure that seems appropriate or that might work as a beginning. Alternatively, if these seem lacking, analyze what is missing, how you would improve the given sequence of information, the wording or order of the various headings.

Style

Style in science generally refers not to the literary qualities of writing, but to the conventions governing its form. In other words, it refers to what is acceptable regarding basic structure, punctuation, capitalization, abbreviations, citation, reference format, and the like. There are a fair number of style guides on the market available in book form (the Council of Science Editors' *Scientific Style and Format* is one example). Most of these are specific to particular fields and are helpful to editors. Many offer necessary advice about using certain universal aspects of style, such as metric notation, conversion factors, various constants, and the like. For everything else, however—that is, the great majority of stylistic aspects—nearly all these guides are of only partial use to the scientist-author.

I make this (heretical) statement for a simple reason: experience. Scientific disciplines are too diverse in their literature to obey a single manual of style. Standards and conventions vary at every level, among journals within a single field, indeed even for single periodicals over time (e.g., between different chief editors). The reasons for this are complex. They have much to do with the interplay between institutional demands and the personalities and training of the people involved.

Editors may well go to war (and the grave) over the proper punctuation in a list or the form of references. But for writers, the practical result is obvious: because no final standards exist, you must take each journal on its own terms, which means examining closely the articles it publishes and, above all, consulting the "instructions to authors" or "suggestions to contributors" requirements for preparing manuscripts. This is the *only* way you can be sure to comply with these basic demands. It points up again the value of using models in science.

Environment

Your work of authorship can be lightened through your choice of setting. By this, I don't mean retiring to a country manor or mountain cabin, however attractive that may be. Instead, I refer here to self-knowledge: be aware of what environmental factors encourage and discourage you from getting to work. "Avoid everyday mediocrity in your working conditions," says Walter Benjamin, one of the great essayists of the 20th century. Some people need a clear desk, nice music (quality sonic wallpaper); others work better if surrounded by notebooks, reports, papers (the voices of data). Some prefer to write in their office, where everything is nearby; others work best at home, with a bit of distance and the motivating nag of family. Learn to know thyself; use your inclinations.

These are not trivial concerns. Writers have a number of tools to put in play; time and place are among them. As a human process, writing has an ecology to it, a personal dimension of engagement that can be nurtured or withered by physical context. The where and when of composition have importance, not just the how and what (we leave the why to the gods of employment and destiny).

Getting Started

If you have trouble facing that blank page or screen in the beginning, here are some techniques to consider.

Try your title first. You probably have some idea of what you're writing about (this would seem necessary). Jot down a possible phrase, even two, or three (or more), to describe your topic, just to start.

Remember what was said about writer's block, near the end of chapter 3? Try some of the ideas listed there to start the flow of language in your mind. Read through an article or two on a similar topic. Be assured: this is *not* wasting time or postponing "the inevitable," but is valuable preparation and may well save time. The more fluent your inner voice, the more easily words will come to you.

Begin the paper with the introduction. If you can't think of an opening sentence, go to your models, choose an article on a similar topic, and imitate—even (to begin with) copy—the first sentence, substituting your own topic and terms. Note how the rest of the paragraph and the remainder of the introduction are structured. Emulate these too, if you find yourself still stuck. Then go back and revise, inserting background information specific to your subject.

Look at a previous paper of your own. How did you begin? See what type of generalization you used. Was it about the importance of the topic? Did you point out a gap in the existing knowledge? Perhaps you simply described the process that you studied or to which you added a new piece of understanding. Briefly analyze what you did and see if it might be useful for your present subject.

Sit down and discuss the subject with a colleague. This will force you to express it and may well plant some phrases in your mind that you can use afterward. Few things help clarify a topic in the early stages more than having to explain it to others. Take note of any questions your associate may have: as a listener, she or he is your first "reader" and can therefore aid you directly in determining what points to cover.

Revising

For Organization

Rewriting is where effective documents are made or lost. It is where good writers and their poorer relatives part company. There is a relatively simple reason for this. Revision is where you, the writer, get the chance to become a reader with the power of change. You see what type of experience the document creates. You see where it is rough and splintery, where it produces discomfort, where it is unfinished. At best, you read through it as if someone else had written it and given it to you to polish up (for a nice fee). You ask, at each step—as every author should—is this material that, if I saw it in print, I would want to have my name associated with it? I'm not saying each sentence needs to be perfect; you're not writing *Madame Bovary* (Flaubert spent a month or more on single sentences and nearly went mercifully insane). But you want to avoid clumsiness, to sound articulate.

So far as I know, there has never been a Mozart of scientific composition. No one ever gets it right the first time—or, for that matter, the second. Revision is the chance we give ourselves to finish a document, to nurture it into maturity. Don't think of it therefore as "making repairs" or "replacing bad with good": rewriting is not how we "fix" but how we *complete* a text. Let's consider an example:

> Reefs of Silurian age have been the main source of oil and gas production in the state of Michigan for the past three decades. In 1990, the reefs produced about 25 million barrels of oil (84% of the state's production) and almost 132 billion cubic feet of gas (92% of the state's production). At the end of 1990, cumulative production from the reefs reached 211 million barrels of oil, 1.21 trillion cubic feet of gas, and 50 million barrels of water. Estimates of the primary recoverable reserves in these reefs are 300–400 million bbls of oil and 3–5 trillion cubic feet of gas. The main purpose of this study is to present a detailed analysis of the depositional history of one of the largest and best sampled Niagaran pinnacle reefs in the Michigan Basin.
>
> Pinnacle reefs of the Michigan Basin are isolated carbonate buildups completely encased by salt, anhydrite, and fine-grained carbonate deposits. Until the 1960s, the gravimetric method was the principal successful tool used for identifying pinnacle reefs in the subsurface. In more recent

years, the search for reefs has been based almost entirely on seismic methods, with exceptionally good results.[1]

We can sense that this is half-formed, both in organization and style. What should be done? It usually helps to look at organization first.

The introductory sentence seems perfectly fine, but it is then immediately swamped by supporting numerical information. After wading through this, we jump suddenly to "the main purpose," which has to do with geologic history, not oil and gas productivity. At the start of the second paragraph, meanwhile, we shift gears again, to a statement about reef character, and, after that, skid into a completely different topic: methods for locating the reefs.

There is a lot of information here, but not all of it appears relevant, and most of it comes at us in disconnected pieces. What is necessary? What can we delete or replace? To answer this, look at "the main purpose" again. If we are interested, above all, in the geologic history of these reefs, do we really need to know how much oil and gas they produced in a particular year, or the particular methods used for finding them? Clearly no. So this tightens things up a bit. Do we want to get rid of *all* the numerical information given in the first paragraph? Probably not, because we need to add support and specificity to the opening claim. Is this first sentence fine the way it stands? Look closely at the rest of the paragraph: we don't learn these are *pinnacle* reefs, specifically, or that we are in the Michigan *Basin* (a particular geologic province), until the last sentence, where these very important terms are merely thrown in. Suppose, then, we revise the first sentence to look like this:

> Pinnacle reefs of Silurian age in the Michigan Basin are a main source of oil and gas, with estimated recoverable reserves of 300–400 million barrels of oil and 3–5 trillion cubic feet of gas.

This keeps the sense of importance that the original opening had, moves all defining terms up to the front, and adds numerical support, all in a rather brief sentence. Using our "main purpose" as an organizing principle, we delete the remaining numbers (we can put these back in a later section)

---

1. D. Gill, "Depositional Facies of Middle Silurian (Niagaran) Pinnacle Reefs, Belle River Mills Gas Field, Michigan Basin, Southeastern Michigan," in *Carbonate Petroleum Reservoirs*, ed. P. O. Roehl and P. W. Choquette (New York: Springer-Verlag, 1985), 123–124.

and skip to a description of the basic geologic character of the reefs—this, after all, helps orient the reader as to what we are talking about.

> Pinnacle reefs of Silurian age in the Michigan Basin are a main source of oil and gas, with estimated recoverable reserves of 300–400 million barrels of oil and 3–5 trillion cubic feet of gas. These reefs consist of isolated carbonate buildups completely encased by salt, anhydrite, and fine-grained carbonate deposits.

Can we now add the last sentence in the first paragraph, to complete our introduction? Let's see if this works:

> Pinnacle reefs of Silurian age in the Michigan Basin are a main source of oil and gas, with estimated recoverable reserves of 300–400 million barrels of oil and 3–5 trillion cubic feet of gas. These reefs consist of isolated carbonate buildups completely encased by salt, anhydrite, and fine-grained carbonate deposits. The main purpose of this study is to present a detailed analysis of the depositional history of one of the largest and best sampled Niagaran pinnacle reefs in the Michigan Basin.

Not bad. But there is still a bit of a jump between the second and third sentences. The paragraph feels too short. And we have sort of tossed out the adjective "Niagaran," hoping that readers will do the work of definition themselves. What needs to be done? How about adding a sentence (shown in bold) to provide transition:

> Pinnacle reefs of Silurian age in the Michigan Basin are a main source of oil and gas, with estimated recoverable reserves of 300–400 million barrels of oil and 3–5 trillion cubic feet of gas. These reefs consist of isolated carbonate buildups completely encased by salt, anhydrite, and fine-grained carbonate deposits. **Previous investigation has established that reefs grew along the basinward margin of a major platform, which rimmed the Michigan Basin in Middle Silurian (Niagaran) time, and were later overlain by evaporite material (salt and anhydrite) due to a major drop in sea level (Hedberg 1975; Corson et al. 1986).** The main purpose of this study is to present a detailed analysis of the depositional history of one of the largest and best sampled Niagaran pinnacle reefs in the basin.

We've taken some information from another part of the paper and put it here, up front. This is definitely better. We've added something that moves the focus from present-day character of the reefs to their geologic history, which is where we want to be to state our main purpose. We've defined "Niagaran," and we've also brought in some references. Notice, too, that our added information partly explains what was given in the previous sentence—in this case, salt and anhydrite are tied to sea-level drop. The transition is thus smooth and logical. Is there anything more we might do? How about making one more segue to the last sentence:

> Pinnacle reefs of Silurian age in the Michigan Basin are a main source of oil and gas, with estimated recoverable reserves of 300–400 million barrels of oil and 3–5 trillion cubic feet of gas. These reefs consist of isolated carbonate buildups completely encased by salt, anhydrite, and fine-grained carbonate deposits. Previous investigation has established that reefs grew along the basinward margin of a major platform, which rimmed the Michigan Basin in Middle Silurian (Niagaran) time, and were later overlain by evaporite material (salt and anhydrite) due to a major drop in sea level (Hedberg 1975; Corson et al. 1986). **Such studies, though valuable, have been largely regional in nature.** The main purpose of this paper is to analyze in detail the depositional history of one of the largest and best sampled individual reefs in the basin, **the Belle River Mills feature.**

At this point, our introduction finally seems complete. We have inserted explanatory transitions that bring the reader from the importance of the subject to the importance of our work. We've moved from the general to the specific, introduced needed terminology. We've revealed the gap in knowledge that our study is going to fill (always a good idea for introductions). We've identified our specific topic and described what we are going to do with it.

So we have a perfect paragraph, at last. Right? Hardly. No such fauna exists. What we have is a piece of writing from which we can move on with a reasonable degree of confidence. If we had world enough and time, we could tinker with our intro a good deal more, ad infinitum. But then there would be the minor problem of never getting it published. Flaubert, of course, did work 20 years on *Madame Bovary*—but novels don't become dated quite so quickly as do data.

Style

I've devoted much space to a single paragraph for three reasons. First, introductions are extremely important, especially to editors and reviewers—if you impress them here, you've done much to get them on your side (so that they may forgive other faults along the way). Second, I've tried to show how all the factors discussed earlier in this chapter come into play: trial and error, experimentation, thinking about your reader, organization. All these factors eventually need to work together. Third, and finally, our all-too-abbreviated example is meant to reveal how much decision making—identifying and solving of problems—actually goes into the rewriting process.

Now let's treat literary style in particular. Note the following passage from a published article:

> As a method to generate low-density microcellular foam, we synthesized molecules that would dissolve in CO_2 under relatively moderate pressures, then associate in solution to form gels. Previous work has shown that gels can be created in traditional organic solvents through hydrogen bonding, association between ionic groups, or association between electron-donating and electron-accepting moieties. To form foams from such gels, it is necessary to preserve the supramolecular aggregates created in solution, both during and after solvent removal.

Right off the bat, we are faced with a dangling modifier ("As a method to . . . , we . . ."). Deleting the first three words provides a cure. Next, in the same sentence, there is confusion created by that last comma and following phrase ("then associate in solution"). At first reading, this seems to relate back to "we synthesized"; on a second or third reading, however, we can see that it doesn't, and that it should form a parallel structure with "dissolve in," as follows:

> To generate low-density microcellular foam, we synthesized molecules that would dissolve in CO_2 under relatively moderate pressures and then associate in solution to form gels.

This reads better. After looking it over several times, however, it seems a bit redundant: if we say the molecules "would dissolve in CO_2", do we also need to say that they subsequently associate "in solution"? This is not a large point, but it allows for extra economy. Moreover, if we wanted to project confidence, as an author, we might also eliminate the conditional tense:

> To generate low-density microcellular foam, we synthesized molecules that dissolve in CO_2 under relatively moderate pressures and then associate to form gels.

That last phrase still hangs a bit. Reading the sentence through, we see that a temporal process is being stated: "that dissolve . . . then associate." What if we make this a bit more explicit:

> To generate low-density microcellular foam, we synthesized molecules that **first** dissolve in CO_2 under relatively moderate pressures and then associate to form gels.

Adding one word makes everything clear at last.

What about the second sentence? Here's a case where enumeration can help:

> Previous work has shown that a gel can be created in traditional organic solvents by one of three processes: (1) hydrogen bonding; (2) association between ionic groups; or (3) association between electron-donating and electron-accepting moieties.

Notice that I've inserted "one of three processes" instead of simply "three processes." As indicated by the final conjunction "or," these processes are not simultaneous or overlapping. This is an example of adding precision and clarity for the reader.

As for the last sentence, a bit of reshaping can be done here too. First, the combination "form foams from such gels" sounds awkward to the ear. What if we changed it to read:

> In order to produce a foam out of such gels, it is necessary to preserve the supramolecular aggregates created in solution, both during and after solvent removal.

Our final problem comes in the last phrase. What, exactly, does it refer to? Does it tell us that aggregates are created both during and after solvent removal? Or does it relate, instead, to the required preservation of these aggregates? As it happens, the rest of the paragraph eventually tells us that the latter is indeed the case, which we might have guessed. But the construction is unclear. Let's therefore change it:

> Creating a foam out of such gels requires that the supramolecular aggregates created in solution be preserved both during and after solvent removal.

This places the verb form "be preserved" closer to its true antecedent "solvent removal" and clears up any confusion. Our final version of the paragraph thus reads:

> To generate low-density microcellular foam, we synthesized molecules that first dissolve in CO_2 under relatively moderate pressures and then associate to form gels. Previous work has shown that a gel can be created in traditional organic solvents by one of three processes: (1) hydrogen bonding; (2) association between ionic groups; or (3) association between electron-donating and electron-accepting moieties. Creating a foam out of such gels requires that the supramolecular aggregates created in solution be preserved both during and after solvent removal.

This is good, functional scientific expression. It is clear enough so that the reader will not stub his or her toe on any broken or confusing phrases. It is reasonably smooth, good in logic, and carries the argument in a particular, desired direction.

It will not have escaped your notice, I presume, that a fair amount of work was needed to get this far. In fact, I have streamlined the process, for the sake of brevity, leaving out a number of dead ends and unsatisfactory changes that I went through in my own revision of the above paragraph. It is important that you know this: perceiving what's wrong with style or organization does not mean that you can automatically correct it without undergoing the process of trial and error.

In the end, these are the sorts of questions you need to ask of your document and then answer through change. If you have any doubts about whether such change is needed in a particular passage, show your article to a colleague (or two)—this is a good idea in any case, particularly after you have "finished" a final first draft. There is no substitute for a foreign pair of eyes. Such foreignness (from the text) is partly what you yourself are after whenever you revise. If this is difficult to achieve, or if you become overly frustrated, put the article aside for a while (a few days, a week, if possible longer), and then return. You may find yourself the prodigal son.

Common Problem Areas

There are, indeed, a very large number of possible stylistic and organizational pitfalls to which we are all prey. A tome of considerable specific gravity would be required to delineate and address even half of them. Here I note only a very few of the most common—those that seem to appear most often in published papers, books, reports, and proposals. Look these over and make sure you avoid being their victim:

THE RUN-ON NOUN PHRASE

Empirical ground-motion seismic attenuation data

or

arctic tropospheric ozone depletion measurements

Stacking adjectives and nouns in this fashion reflects an attempt at increased shorthand, but it produces literary plaque and turns any sentence or paragraph varicose. Though scientific language is dense, it cannot do away with readability altogether. Try, instead,

empirical data on seismic attenuation from ground motion studies

and

measurements of ozone depletion in the arctic troposphere

If, however, the phrase is an accepted (standard) technical term, use an abbreviation, for example,

high-mobility Si metal oxide semiconductor field-effect transistors (MOSFETs)

THE MISPLACED DEFINITION

Strong ground motion information (estimates of peak amplitudes, durations, and phasing of seismic waves) is necessary to engineer earthquake-resistant buildings.

Definitions are best placed either at the end of a sentence or in a separate sentence. Otherwise, they tend to jam or disperse the flow of reading. Try, instead,

> To engineer earthquake-resistant buildings requires strong ground-motion information, that is, estimates of peak amplitudes, durations, and phasing of seismic waves.

or

> Information on strong ground-motion is necessary to engineer earthquake-resistant buildings. Such information includes estimates of peak amplitudes, durations, and phasing of seismic waves.

DISJOINTED TOPICS

> Increasing global temperatures have been predicted to shift the geographic ranges of many species to higher latitudes or altitudes. Such increases in temperature have been tied to changing chemical composition in the troposphere, particularly the influx of carbon dioxide from combustion of organic fuels. Here we report on experimental evidence that slight fluctuations in water temperature control the impact of a key predator, the starfish *Pisaster ochraceus*. Preliminary analysis of *Pisaster* behavior indicates that it becomes inactive in low zone channels during periods of upwelling.

If the first sentence presents a good opening, the second jumps to a completely different topic. It should be deleted here and saved for a separate section, most likely the conclusions. The third sentence, meanwhile, relates to the first, but the last sentence presents another leap to a new topic and should be placed elsewhere, possibly in the following paragraph. Try, instead,

> Increasing global temperatures have been predicted to shift the geographic ranges of many species to higher latitudes or altitudes. Locally, communities may undergo changes in composition as species adapted to warmer temperatures are progressively introduced from lower latitudes. Here we report on experimental evidence that slight fluctuations in water temperature control the introduction of a key predator, the starfish *Pisaster ochraceus*.

In all cases, place similar topics or types of data together. Avoid "dumping" information on the page in the hope that the reader will make any necessary connections and thus do the hard work for you. Good organization involves grouping and ordering material. Go back and read some of your models with this in mind: see if they measure up.

THE LAND-RUSH OPENING

Intense bursts of gamma rays that last up to several minutes have been observed during the past three decades and produce an optical afterglow believed to be synchrotron radiation from an expanding ultrarelativistic blast wave whose exact source geometry and emission mechanism is unknown but can be constrained by accurate optical polarization measurements.

Now take a breath. In good, functional scientific writing, the opening paragraph is divided into two basic parts: a short first sentence, which acts like an overture, and a series of following sentences, usually longer, that develop the topic into more specific variations. For the above, one might start with

Intense bursts of gamma rays that last up to several minutes have been observed by many astronomers during the past three decades.

Or place it in the active voice:

During the past three decades, many astronomers have observed intense bursts of gamma rays that last up to several minutes.

Then follow with

When an optical afterglow is observed following such gamma-ray bursts (GRBs), they are interpreted to be synchrotron radiation from an expanding ultrarelativistic blast wave. The exact source geometry and emission mechanism for this blast wave remain unknown, but can be constrained by accurate optical polarization measurements.

The opening paragraph should *introduce*—it needs to usher readers into the subject, not inundate them with a shower of topics and terminology.

Move from the general to the specific, present key words in stepwise fashion, bring readers up to speed.

Process

What of practical methods or techniques in revising? The advent of computers and word processing has changed the process of rewriting forever. It is now possible to revise as we go along, leaving no tracks behind us. Technology has helped eliminate any obvious boundaries between writing and rewriting.

A very helpful procedure, particularly if you find it difficult to get started each time you sit down, is to read through what you've already done, get the material flowing in your head, then extend it forward. As you do this, you will probably find things to change before you get to your stopping point. Most people do revise every time they write. Something, somewhere, begs to be changed. This might well result in the earlier portions of your document being more polished than the later ones: try to be aware of this and to note if any important discrepancies in style exist.

To revise properly, you must be able to see your own work from a distance. This is where your models can come in again. Before going through your document to make changes, choose an article from your collection of good writing, preferably on a topic related to your own, and read it to get a sense of sound and style. You may need to do this more than once. Then look through your own work. How does it compare? Is it fairly smooth, logical? Where are there "bumps" or "pot holes"? Where is it well done? What can you do to make parts of it read more like your model or your higher-quality sections? Using your models of good expression as a measuring stick can be very useful in pointing up instances where you need to revise and, also, those where you don't. If the first will keep you working, the second will help give you the confidence to do so. Don't be shy about admitting your successes, even if they involve a single sentence here and there. Good writing is always an admirable achievement, no matter how locally it occurs.

✳

5. WRITING VERY WELL: OPPORTUNITIES FOR CREATIVITY AND ELEGANCE

Few things are impossible to diligence and skill. —SAMUEL JOHNSON

Beyond the Functional: "Creative Writing" in Science

One of the most unfortunate folktales about science is that it has no room in its daily expression for creativity and personal eloquence. Except, perhaps, for a lecture by one or another towering genius or a popular account by a researcher who, to the amazement of all, can "write excellent prose for the layman," scientists are presumed to inhabit a world utterly lacking in literary atmosphere, a planet not merely flat and functional, but bleak, distant, and cold. Indeed, the worst aspect to this bit of folklore is that scientists tend to believe it themselves.

The truth, however, lies elsewhere. All forms of persuasive writing can be shaped creatively, with dashes of refinement, even beauty. Does this actually happen in science? Yes, indeed. Expressive distinction has not entirely escaped the gravity of the "scientific," by any means. It may be relatively rare—one article in a hundred, a few paragraphs in a lengthy review paper, scattered sentences in otherwise ordinary articles. But it is there, a figure in the carpet. Some observers have noted the fact that the famed paper on DNA structure by Watson and Crick (analyzed in chapter 2) has been called "a prose poem" by more than a few scientists. And though this kind of praise may well exaggerate the case, it still shows that elegance does not go unappreciated.

What does it mean to write elegantly, creatively, in science? Basically, it means creating texts that are not merely effective but *interesting* to read, those that offer a higher order of reading experience. In practical terms, this means fulfilling all the requirements of good, functional communication in a manner that is rhetorically sophisticated, even inventive. There are many techniques that graceful authors employ, either consciously or intuitively. They include, for example, modulating sentence length and rhythm, posing questions at particular points in the narrative, applying parallel structure in individual sentences, employing refined turns of phrase, coining new terminology, providing smooth transitions between paragraphs and sections, and using syntax as an echo to meaning.

Such techniques represent a high level of expressive skill no matter what field or subject is involved. But in science, special circumstances need to be considered. As writers, scientists work within a system of expressive restraint and so must be relatively subtle in their creative acts. Conventions dictate that they not show off, that they avoid drawing too much attention to their language as an aesthetic medium. In the average technical paper or report, there are particular opportunities for applying elegance and invention. Let's look at some of them.

Opportunities

Occasions for elegance occur in many parts of a document. Of course, any article or report can be made more graceful by excellent organization and polished writing throughout. But there are also particular points in an argument where good writing can itself be raised to a higher level with specific purpose. Here are some of them:

1. *Any place where there is a need for generalization.* This is a chance for adding memorable or inventive phrasing. Examples include the introductory and concluding sections of a document, especially in the very first and last paragraphs.
2. *Similarly, the first and last portions of each individual section.* This provides an opportunity for generalizing gracefully and for providing subtle transitions, for example, those that pick up on points previously made or that forecast or suggest what comes next.
3. *Points in the argument that call for an added degree of emphasis.* This may involve introducing an important, even unexpected, result; revealing a

gap in the existing literature; stating a weakness in current theory; pointing out the (magnificent, unparalleled . . .) contribution of your work; suggesting areas for future investigation.

4. *Places where the argument takes a new turn,* for example, where it embarks on a new topic or direction, a different area of data or type of measurement.

5. *Areas in a paper or report where the language is descriptive and less technical,* for instance, where some aspect of history (e.g., of the existing literature or a particular technique) is discussed.

6. *The last sentence.* Here is where readers leave your text, thus where you can leave them with a memorable and meaningful line, elevating your paper above others.

This list is meant to be illustrative, not inclusive. Different authors have often created different chances to assert their originality. What follows, then, is a series of examples aimed at giving instances applicable to all of the above opportunities, plus others as well.

Example 1: Introduction and Conclusion

Note this beginning to a paper entitled "Labyrinthine Pattern Formation in Magnetic Fluids":

> Several distinct physical systems form strikingly similar labyrinthine structures. These include thin magnetic films, amphiphilic "Langmuir" monolayers, and type I superconductors in magnetic fields. Similarities between the energetics of these systems suggest a common mechanism for pattern formation.[1]

This is a highly technical opening, yet with certain raw ingredients for eloquence. The first sentence is short and dramatic, the second follows by introducing specific topics, and the third reveals the fundamental logic underlying the investigation. Three brief sentences do a great deal of necessary work, in admirable order. Let's work with them a bit:

> Different magnetic fluids can form strikingly similar labyrinthine structures. In cases where similarity of structure reflects similarity in fluid en-

---

1. A. J. Dickstein, S. Erramilli, R. E. Goldstein, D. P. Jackson, and S. A. Langer, "Labyrinthine Pattern Formation in Magnetic Fluids," *Science* 261, no. 5124 (1999): 1012.

ergetics, a common mechanism may exist for pattern formation. Three examples for which this assumption holds true include thin magnetic films, amphiphilic "Langmuir" monolayers, and type I superconductors in magnetic fields.

Analyze what has been done to the passage. Note the changes in sound (e.g., deletion of too many *s* words in the first sentence), the added parallel structure ("similarity of . . . similarity in . . ."), and the new (previously hidden) logic in the second sentence. Note, too, sentence length—short, medium, long—which acts to draw the reader in and, at the same time, allows the introduction of increasing detail. Here, meanwhile, is the conclusion to the same paper:

> In conclusion, the branched patterns seen in experiment can be understood within perhaps the simplest dynamical models incorporating the competition between surface and dipolar energies. Taken together, the experimental and theoretical results indicate that an enormously complex energy landscape in the space of shapes can arise from a competition between short-range forces and long-range dipolar interactions in systems subject to a geometric constraint. A static theory of labyrinths would find only the minimum energy configurations, whereas the dynamic theories reflect the complexity of the landscape in the complexity of the labyrinths.[2]

Again, we see the impulse toward eloquence, the desire for, and partial achievement of, making memorable statements. Let us, therefore, help things along:

> To conclude, we propose that the branched patterns seen in experiment be understood within quite simple dynamical models. Such models, in contrast to static theories, will be able to incorporate the observed competition between surface and dipolar energies. Taken together, both the experimental and theoretical results indicate that, in systems made subject to a geometric constraint, an enormously complex energy landscape may arise from this contest between short-range surface forces and long-range dipolar interactions. Static theories cannot possibly account for

2. Ibid., 1015.

such competition, but derive only the minimum energy configurations required for labyrinths to form. Dynamic theories are needed to reveal how complexity of the energy landscape is reflected in the complexity of labyrinth structure.

Once more, look at what was done. Very little by way of content was added or deleted. Instead, certain elements were reordered; sentences were divided up to give each individual point its own emphasis. A word here and there (e.g., "quite," "contest," "possibly") is employed for effect, without taking anything away from the content. Note, too, the passage is a bit longer than the original and, to purists, may even seem a bit redundant. An elegant conclusion, however, produces an effect that sticks with the reader: if perused, the above will reveal no repetition, but instead a degree of overlap that helps drive the central argument home. The paragraph ends, meanwhile, with what can only be described as a transparently veiled philosophical statement, a neatly packaged "grand principle" that all scientists would recognize as one of their own. Is there any reason to hold back from this sort of finale? Absolutely not—as long as you do it in the specific terms of your subject.

Example 2: Reorganization and Polish

Recall the following paragraph, part of which was examined in chapter 3 as an example of high-quality writing:

Subduction of the Juan de Fuca and Gorda plates has presented earth scientists with a dilemma. Despite compelling evidence of active plate convergence, subduction on the Cascadia zone has often been viewed as a relatively benign tectonic process. There is no deep oceanic trench off the coast; there is no extensive Benioff-Wadati seismicity zone; and most puzzling of all, there have not been any historic low-angle thrust earthquakes between the continental and subducted plates. The two simplest interpretations of these observations are: (1) the Cascadia subduction zone is completely decoupled and subduction is occurring aseismically, or (2) the Cascadia subduction zone is uniformly locked and storing elastic energy to be released in future great earthquakes. Full resolution of this issue may prove elusive. Although it is somewhat surprising that no

shallow subduction earthquakes have been documented in this region,
the duration of written history is relatively short.[3]

As it happens, the paper does not begin with this elegant paragraph. Instead, it starts: "This is the third in a series of four papers that lead to an es-
timation of the seismic hazard associated with the subduction . . . beneath
North America." The passage above follows third, after a discussion of the
existing literature and the key aims of the article. To revise with polish, we
would reorganize the introduction to begin with the above, follow with
discussion of the literature, and end with mention of the paper's goals.

Can the above paragraph, meantime, be improved? Let's try.

> Subduction of the Juan de Fuca and Gorda plates has presented earth sci-
> entists with a dilemma. Despite compelling evidence of active plate con-
> vergence, subduction on the Cascadia zone appears to be a relatively be-
> nign tectonic process. There is no deep oceanic trench off the coast; there
> is no extensive Benioff-Wadati seismicity zone; and most puzzling of all,
> there have not been any historic low-angle thrust earthquakes between
> the continental and oceanic plates. The two simplest interpretations of
> these observations are (1) the Cascadia zone is completely decoupled and
> subduction is occurring aseismically, or (2) the Cascadia zone is uniformly
> locked and storing elastic energy to be released in future great earth-
> quakes. Although it is somewhat surprising that no shallow subduction
> earthquakes have been documented in this region, the duration of written
> history is relatively short. Full resolution of this issue may prove elusive.

In this case, we have done very little by way of revision. The word "subduction" or "subducted" was replaced in a few instances to reduce over-
repetition of terms (very common in scientific writing). In the second
sentence, the phrase "has often been viewed as" was changed to "appears
to be," since, from the foregoing sentence, it is the process of subduction,
not views of it, that forms the stated dilemma. The final two sentences

---

3.   T. H. Heaton and S. H. Hartzell, "Source Characteristics of Hypothetical Subduction
Earthquakes in the Northwestern United States," *Bulletin of the Seismological Society of America* 76, no. 3 (1986): 675–676. In the succeeding examples, I have adapted parts of the article
by Heaton and Hartzell to my purposes, revising the actual sections to varying degrees. Despite this treatment, it should be clear that this is generally a good example of an eloquent
paper in the earth sciences.

have been reversed: this is not only good for logic, but also for sound and rhythm. Graceful opening and closing paragraphs usually begin and end with short sentences. On the other hand, nothing was done to the eloquent, even oratorical third sentence, with its wonderful parallel structure. Note how effectively and concisely the three points are delivered (three is the magic number here, stemming from an ancient rhetorical formula, as in Lincoln's famous "of the people, by the people, for the people").

Example 3: Use of Questions

Use of questions is a simple literary technique only sometimes used in science, but it can be a graceful and effective accent that serves to add emphasis, introduce new material, or refocus a discussion. Posing a question adds a bit of drama:

> Cholesterol and related sterols are not uniformly distributed within the membranes of eukaryotic cells. Why is this so? Here we seek an answer by considering the effects of these flat, disc-like molecules on lipid bilayers.

A question can orient and hold the reader's attention, break up an extended discussion, or turn the narrative in a particular direction:

> It has been observed that annual cycles of measles epidemics occur in areas where the birth rate is high. What are the limits to such a correlation and how can they be established? Previous models have assumed an average, therefore constant, birth rate, ignoring any effects that might result from seasonal variations.

There are also more restrained forms of asking questions, if the more overt query seems too bold for your particular publication venue:

> Measurement of ultrashort pulses is a demanding task. The question has been asked: how can our existing instrumentation be improved? Our results suggest that . . .

Or, in an even more subtle style:

Recent climatic modeling has revealed that maximum flux of anthropogenic carbon into the Northern Ocean occurs farther north than previous inventories would suggest. To investigate why this might be the case, we examined two mechanisms of transport.

The onset of AIDS appears associated with an extension of HIV infection beyond the lymphoid organs into tissues such as the brain. Just how this occurs is unclear, but it seems probable that infected cells are among the responsible agents.

Note that the posing of questions does not have to be restricted to the opening portions of a document. It can also be appropriate in the main body of your discussion,

During the analysis of experimental results, the question arose: what mechanisms might account for the single-file transfer of particles in this setting?

your conclusions,

Is slip along the Cascadia subduction zone a benign process, occurring slowly as aseismic creep? Or alternatively, is elastic strain energy accumulating along this zone, and if it is, what is the nature of earthquakes that may result?

or even headings,

Cascadia Subduction Zone: Locked or Unlocked?

Single-File Particle Transfer: Where and How Does It Occur?

Inserting questions into your document should be done sparingly. Queries of this type are forceful signposts for the reader: use them too often and you risk making your audience feel they are being manipulated (a great failure for any author). Place your questions in carefully chosen locations. As always, if you're experimenting with this technique and are somewhat unsure of its results, have a colleague read what you've written. Better yet, have two or three read it.

Example 4: Transitions

One of the distinguishing features of any written document lies in the elegance of its transitions—how it knits together its various pieces, establishing a smooth flow between them. A simple technique for doing this is to announce, at the end of your introduction, what topics you will be discussing and then to proceed to do this in the same order mentioned, with each topic included in the headings of the following sections.

For example, the paper mentioned above on the Cascadia subduction zone and its potential for earthquake activity ends its introductory opening something like this:

> In this report, we extend the work of previous authors by systematically comparing trench bathymetry, gravity, and shallow seismicity for a worldwide sampling of subduction zones. It is hoped that by pursuing this line of investigation, one or more analogs may emerge for consideration.

The sections of the report that follow would ideally include headings such as Trench Bathymetry and Gravity; Seismicity; Specific Comparisons; Analogs: Do They Exist?; Earthquake Potential along the Cascadia Subduction Zone; Conclusions.

Such linkage is needed in functional writing too, of course. Elegance enters into the equation in the logic of the general plan, but also in transitions that are given *between* individual sections. For instance, the first section (Trench Bathymetry and Gravity) could begin with the sentences

> The Cascadia subduction zone is unusual in that it has virtually no bathymetric trench. To begin to assess just how anomalous this is, we have constructed profiles of bathymetry and free-air gravity for many circum-Pacific convergent boundaries.

and end with

> We note, in particular, striking similarities in both bathymetry and gravity profiles between Cascadia and the subduction zones of Colombia and southern Chile. This raises the question of seismic activity.

These sentences obviously pick up on the opening, while providing flow into the next section (Seismicity). But note too that they provide a first an-

swer to the opening query—how anomalous is Cascadia—and thus establish a question-answer pattern that might be pursued in each succeeding
part of the paper. Indeed, the next section could then begin like this:

> A second apparent anomaly associated with the Cascadia subduction
> zone is the remarkable paucity of shallow earthquakes. While this seems
> puzzling for a major plate boundary that is undergoing 3–4 cm/yr of con-
> vergence, it bears asking whether and to what degree analogous condi-
> tions may exist elsewhere.

Note how several levels of transition are employed. First, we have carried
forward the idea of (and the word) "anomaly" presented in the opening of
the earlier section. Second, we now propose a possible contrast term, "ana-
log," implicit in the final sentences of the present section and mentioned in
the last part of the introduction. Third, we emulate—without copying, but
instead by adding variation to—the question-answer pattern mentioned
above. Notice that Colombia and Chile, as possible analogs, are not mentioned. On the one hand, the relevant similarities must be established in
relation to worldwide comparisons (this is simply good science). On the
other hand, however, keeping them out here leaves the reader with a slight
edge of anticipation, a heightened interest: will these same areas rise again
out of the mix? Proceed on, gentle reader . . .

Example 5: Word Choice

Choosing specific words is one area where inventiveness can be employed
in an immediate way. Elegant writers consciously vary their vocabulary,
trim repetition. At times, they will seek an unusual or unexpected word to
create added interest. Look at one of the previous examples and a rewrite
of it (bold type indicates changed words and phrases):

> In conclusion, the branched patterns seen in experiment can be under-
> stood within perhaps the simplest dynamical models **incorporating** the
> competition between surface and dipolar energies. Taken together, the
> experimental and theoretical results indicate that an **enormously com-
> plex energy landscape** in the space of shapes can arise from a **competi-
> tion** between short-range forces and long-range dipolar interactions in
> systems subject to a geometric constraint.

In conclusion, the branched patterns seen in experiment can be understood within simple dynamical models that **embrace** competition between surface and dipolar energies. Taken together, both the experimental and theoretical results indicate that, in systems made subject to a geometric constraint, an **energy landscape of enormous complexity** may arise from this **contest** between short-range surface forces and long-range dipolar interactions.

Only a few words have been replaced, yet they give the passage a new elegance. The word "embrace" is fully acceptable and certainly no more ambiguous than "incorporate," but much more interesting and suggestive. Meanwhile, "contest" helps us avoid repeating "competition" and adds something extra, a note of struggle.

Here's another example:

A central tenet of biomineralization is that the nucleation, growth, morphology, and aggregation of the inorganic crystals are regulated by organized assemblies of organic macromolecules. Control over the crystallochemical properties of the biomineral is achieved by specific processes involving molecular recognition at inorganic-organic interfaces.[4]

This is fairly clear, functional writing. We can appreciate here the use of the word "tenet" instead of the more common "concept" or "principle." But the rest of the sentence needs a little help, mainly because of word choice and word organization. Try this:

A central tenet of biomineralization holds that the nucleation and growth of inorganic crystals, as well as their morphology and aggregation, are governed by structured assemblies of organic macromolecules. For a particular biomineral, crystallochemical control depends on processes that involve molecular recognition along inorganic-organic interfaces.

Compare the differences closely. Nothing is lost in the change, except perhaps the attempted parallel structure between "inorganic crystals" and "organic macromolecules" in the first sentence, which did not work. We have replaced a number of ordinary words of general meaning ("is," "regulated," "achieved") with more graceful equivalents ("holds," "governed," "depends

---

4.   S. Mann, D. D. Archibald, J. M. Didymus, T. Douglas, B. R. Heywood, F. C. Meldrum, and N. J. Reeves, "Crystallization at Inorganic-Organic Interfaces: Biominerals and Biomimetic Synthesis," *Science* 251, no. 5126 (1993): 1286.

on"), and have shortened a five-word phrase down to a two-word alliterative ("crystallochemical control").

Finally, very often in scientific writing, we feel the need to repeat technical terms, almost to the degree of absurdity, because there appear to be no alternatives:

> Fracture analysis endeavors to measure the spacing and aperture of individual fractures, frequency of fracture occurrence, and the total extent of the fracture network.

This type of writing is, at best, barely functional. It is equivalent to a form of terminological dumping. Try, instead,

> Fracture analysis is an attempt to measure the spacing, aperture, and frequency of fracture occurrence, as well as the dimensions of the relevant structural network.

Adding elegance through word choice can also involve deleting certain unnecessary elements, cleaning up, in other words.

Example 6: Phrasing

Refined phrasing in science? But there are many opportunities and much evidence for this. In fact, with its articulated demands for brevity and concision, scientific writing is one of the very *best* places for the felicitous turn of phrase. Note that the classical aphorism or maxim—a much admired literary model through the ages—derived much of its elegance from placing complex thoughts into highly confined form ("As the scale bends to a weight, so must a balanced man yield to circumstance" [Cicero]). While our aim in science is not necessarily to set the literature aglitter with gems, we should recognize that creative phrasing is a common option, at times even with a drop of humor.

There are no final techniques for this type of writing that can be relied on—except for emulating the work of others, of course. Here are but a few examples that I have collected:

1. A dearth of direct evidence urged us to search for the missing parameters.
2. Gravity waves in the lower atmosphere, forced by flow over mountains, have been observed and modeled for many years.

3. The role of the transfer matrix t is merely to modify this balance quantitatively as long as the lattice is perfectly periodic.

4. Blood vessels are the life-giving conduits that connect our tissues and our organs.

5. Research on thin polymer films has proven to be a drama of refound opportunity.

6. Multiple schemes for single measurements have too often yielded perplexing results.

7. During the years of its early development, the technique of synchronous laser pumping was advanced by a scattering of research teams.

8. The free energy of any physical system is rarely if ever free, but must instead be liberated by one or more conditions.

Notice in these phrases the use of sound and rhythm (examples 1–3), the artistic compression and neatness of word choice (4–6), and the employment of humor (7–8).

Such writing is excellently used as an occasional spice to a document, giving it flavor and finesse, raising it above the level of the ordinary. It is easy to overdo, however. Being too clever, or clever too often, will trivialize your subject. The best writers know to leave the reader affected and wanting more. To satiate completely, as Seneca says, is to erase an impression. A few, well-placed pearls will be more than sufficient for any single text.

Example 7: Metaphor

Contrary to what is often said, and far too often advised, the use of metaphor is alive and well in scientific composition. Scientific terminology literally teems with metaphoric terminology: white dwarf star, killer T-cell, RNA editing, plate tectonics, quantum charm, reporter genes, structural relaxation, molecular target, and so forth. True, these terms only begin as metaphors, when first introduced. Over time, with standardized usage, they lose this charge and become identified with the phenomena they represent (i.e., they are no longer figures of speech in the active sense). But what they reveal, without question, is how strong and accepted the metaphorical impulse remains in science, indeed how *necessary* it is.

In what ways can this be applied to writing, in the ordinary sense? In fact, such application happens all the time, though usually in subtle fashion. Note this example once more (in its original form):

In conclusion, the branched patterns seen in experiment can be understood within perhaps the simplest dynamical models incorporating the **competition** between surface and dipolar energies. Taken together, the experimental and theoretical results indicate that an enormously complex **energy landscape**...

Here are some others:

Synapses are **focal points** of communication between nerve cells.

One of the most dramatic events in the fossil record is the **explosive** diversification of marine invertebrates early in the Cambrian period.

It was soon realized that determination of the exposure history of individual grains was complicated by regolith processes, namely, meteoroid impact "**gardening**" on the lunar surface.

With a homogeneous thermal boundary condition, convection . . . consists of nearly two-dimensional turbulence, with **meandering plumes** . . . . The mean zonal flow is westward, and the convective pattern **drifts** westward with this flow.

After reading these examples, look through your models and see where similar uses of figurative language appear. You'll note that it doesn't happen very often, only at particular points in an argument or document. On the other hand, for any given text, it may not happen at all—if the needed conditions do not arise.

Metaphors tend to enter scientific speech when authors reach for a descriptive word outside the normal pale of technical terminology. In some cases, this is done to fill a gap; in other cases, its purpose (consciously or otherwise) seems more imaginative and aimed at making the text more interesting. Writers might freely adopt a term from elsewhere in their discipline, or they might take it from another discipline, or even from ordinary speech. All of these approaches are entirely acceptable, provided that the chosen word or phrase is appropriate to the case.

In the above examples, figurative language appears in the form of both nouns and modifiers. One case (the third sentence) attempts (it seems) to coin a new term—note how the quotation marks serve to qualify the attempt, but also to let the reader know that the writer is being consciously inventive. New terms, of course, are being proposed all the time in science;

this is both a necessity, as new phenomena are revealed, and an impulse, as individuals try to make a mark on their field. If you are lucky enough to be in a position to coin a new term, please do so thoughtfully, with a degree of intelligence (as a negative example, using comic-strip characters to name planetary features has the effect of trivializing). Such instances are revealing: they show us places where the wider culture enters into science in a direct way, and therefore that scientific speech is hardly as flat and instrumental as is so often maintained.

Example 8: Alliteration

Simply defined, alliteration is the repetition of a letter or sound at the beginning of adjacent or neighboring words. "Nattering nabobs of negativism," is an example from Richard Nixon's vice president Spiro Agnew, who directed these terms of endearment at students marching against the Vietnam War. In truth, alliteration is a stylistic device often elegantly employed by poets, elderly or young, and by literary lions of prose. Yet it also appears in many sayings—"a house is not a home," "forewarned is forearmed," "dead men tell no tales" (two separate alliterations here, in the "eh" sound of "dead" and "men" and in the repeated *t*)—and, incidentally, even in idle, ordinary speech, for example, "What a waste of water!"

But for high-level writing, it needs to be wielded intentionally. By now you know not to ask if scientific expression allows for such things. So let's go right to some examples.

> Analysis of archaeal genome sequences has advanced rapidly since associated DNA replication studies first appeared.

> Among the least-altered objects surviving from the origin of the solar system, comets are commonly understood to carry especially revealing components.

Notice in the first example that the sentence reads smoothly and would probably not have its alliterative dimension noticed by many readers. Yet the repeated short *a* sound—appearing not only in the first letter, but also in "has" and "rapidly"—creates a subtle kind of rhythm that flows through the sentence and makes it move forward more easily than if it were written like this:

Archaeal genome analysis has progressed rapidly due to concomitant DNA replication studies.

Word choice matters, and having an elegant device in your mind when you make such choices can raise your writing to another level. Beware, however: the secret to using alliteration in a classy way is to make it seem natural, unobtrusive. Suppose, for instance, in the second sentence above, we made a few small changes:

Among the most unaltered objects from the origin of the solar system, comets are commonly understood to carry key components.

Suddenly, as it were, the veil is yanked away, the writer stands naked before us, working his or her devices with manipulating intent. Rather than a touch of the refined, we are treated to an overdone effort by an amateur. As I have said before, reading your work out loud can help avoid such clumsiness.

At the highest level, meanwhile, alliterative choices have real substance as well as style. They can go so far as to echo in sound what is said for meaning, or provide it with a degree of dramatic effect.

This report pays special attention to the growing mass of plastic pollution that now plagues the world's oceans.

Repeated use of *p* adds an assertive note of urgency and even accusation, like a finger tapping the table. Again, notice that even a single word change, "growing" to "proliferating" or "propagating," would again be too much and, for many readers, would even trivialize the point being made by suggesting that the writer thinks effects are more important than content.

Example 9: The Last Line

Every skilled writer knows the value of an effective closing. At best, it is like a silent explosion that takes the reader's breath away yet leaves his or her mind saturated with meaning. There are many examples from the history of modern science that achieve, or come marvelously close to achieving, this. One of these we have seen already, it being Watson and Crick's wry and understated last line: "It has not escaped our notice that the specific pairing we have postulated immediately suggests a possible copying mechanism for the genetic material." Another, longer finale, resembling the close of a great Victorian novel comes from Darwin's *Origin of Species*:

> There is grandeur in this view of life, with its several powers, having been originally breathed into a few forms or into one; and that, whilst this planet has gone cycling on according to the fixed law of gravity, from so simple a beginning endless forms most beautiful and most wonderful have been, and are being, evolved.[5]

Today's journal editors are perhaps less inclined toward such grand endings. They are, however, fully amenable to such elegant last lines as these:

> Understanding the detailed dynamics of such a system remains a formidable challenge, which may well demand the development of new theoretical techniques.

The repetition of *d* and the unexpected word "formidable" take this closing to a higher level. Now consider this:

> In addition, understanding the role of mechanics in these biologically inspired designs may help engineers to develop seahorse-inspired technologies for a variety of applications in robotics, defense systems, or biomedicine.

Rewriting the example in a more interesting way might yield something like this:

> Inspired by the biomechanics of these designs, engineers might find in the seahorse a basis for novel technologies in areas as varied as robotics, defense, and biomedicine.

---

5. This ending appears in the original 1859 edition. The foregoing sentence may help us appreciate Darwin's finish a bit better: "Thus, from the war of nature, from famine and death, the most exalted which we are capable of conceiving, namely, the production of the higher animals, directly follows." Online versions of all six editions of the book (plus all of Darwin's other writings) can be found at Darwin Online (http://darwin-online.org.uk/).

As it is the last sentence in the article, we don't need the transition "in addition." We should also take out one "inspired" and grant the engineers the active tense, since they are the ones called on to do something. Another "small" change is to turn "defense systems" into the one-word "defense" so that it matches the others in the list.

Does all of this seem like much ado about too little? Yet the larger point is that such choices can become instinctive with a bit of time and practice, so that all of your writing rises in quality, becoming more concise and readable, earning points with editors and readers. Moreover, this kind of skill is transferrable to other types of writing you might do, making you a better communicator, period. No guarantees that this will happen quickly. Only that if no effort is made, no change will occur.

Another example:

> In remote mountain habitats—largely isolated from habitat destruction, toxins, and pathogens—evolution is helping wild bees keep pace with climate change.

This counts as a fairly decent ending. With a few changes, however, it could be more. Note that having a key phrase jailed between dashes doesn't seem so useful. In fact, rhetorically speaking, it draws attention away from the last part of the sentence, which is really the main point but comes across as emotionally flat. A few alterations can help improve things. I offer two new versions:

> In remote mountain areas, isolated from toxins, pathogens, and habitat destruction, evolution has endowed wild bees with adaptations that keep pace with the effects of climate change.

> Wild bees in remote mountain areas, free from toxins and pathogens and habitat destruction, have undergone evolutionary changes that keep pace with the growing impacts of global warming.

Both of these rewrites have advantages. The first is more restrained but still has more force than the original. The second rewrite introduces several modifications that increase the effect still further, while remaining within the normal limits of scientific discourse. Shifting the bees to the subject of the sentence helps this impression, as does the addition of another "and" after "pathogens" and the words "free" and "growing."

These two versions should help clarify something. In any of the cases above, there is no final, ideal version we should discover or aim at. There are always alternatives to be found, some better than others, some just as good. The written word is highly flexible, even with only a few moving parts, and is all the more so in the hands of a skilled artisan. Recall that the great French author Gustave Flaubert famously spent weeks and even months on crafting a single sentence. He is considered a great master of style, but the price he paid in time and self-torture seems a model for very few, including no scientific writers!

In the End: Literary Finesse Is Knowing When to Stop

Just a brief final note. In nearly all cases, true authorial expertise in science lies in subtlety and restraint, not showmanship. There is certainly room here and there for play, whether this involves chiseling a suggestive phrase or coining a clever term. But on the whole, these opportunities are relatively rare. Science is reserved in its discourse; this is a historical condition, as I've said (chapter 2), but also a kind of knightly code.

Several times in the foregoing sections I've stated the importance of using one or another form of elegant writing on a selective basis. Let me here emphasize this once more. If you find yourself drawn to some of the techniques mentioned above and wish to experiment with them in a document you're working on, by all means do so with multiple blessings, but try to be conscious of the effects you're creating, how often you're creating them, and how obvious you make them. Remember, too, that anything you produce will eventually have to pass through the editorial gate, which can be quite narrow. If you have any doubts, show it to two or more colleagues and note their response (you may have to weigh this somewhat against what you know of their own literary inclinations and limitations). Editors and reviewers usually begin to shudder at the point where they feel "science" is being sacrificed to "style." The challenge for the sophisticated writer is always to make his or her text a stylistic achievement, while, at most, drawing only momentary or background attention to the fact. Elegance and restraint share bread, on the page as in the lab.

＊

6. THE REVIEW PROCESS: DEALING WITH CONTENTS AND DISCONTENTS

Your work is both good and original. Unfortunately the parts that are good are not original, and the parts that are original are not good.—ATTRIBUTED TO SAMUEL JOHNSON

Editors and What They Do

Editors have one of the most difficult, thankless, and important jobs in all of science. The journal editor, in particular, is usually a scientist (one of us) and a volunteer, who gets no remuneration for his or her efforts, but who must keep up his or her own research, teaching, corporate responsibilities, and publishing attempts without skipping a beat, while somehow taking on the added jobs of literary manager, quality control expert, and guardian of the field. This is the situation on the professional level. On the human level, meanwhile, the editor's work involves trying to minimize the indiscretions of others and, therefore, casting the stones of failure and success, hurt and happiness, on every side, all in the name of better science. Which helps explain why the editorial profession in science, like certain gasses, is both noble and volatile.

Another point should be admitted. The strength and direction of a scientific field largely rise and fall with the quality of its editors. Bluntly put, editors have power and influence. Their specific work is to act not only as gatekeepers but as architects, to determine how wide or narrow a range of subjects will be accepted, which incoming articles are worthy of review, who will review them, what range of comments are to be made, whether these comments are to be accepted on face value, and what the final decision will be regarding acceptance, acceptance with major revision, or rejection. These are all critical steps in determining what science is published. Well-written articles on topics of significant interest are a mark of success, both for the authors and the editors. But the opposite is at least as true.

In most cases, an editor is guided in his or her work by certain protocols. For example, journals commonly have a bank of associate editors responsible for individual subfields, and to these people the head editor will delegate the responsibility for handling many of the papers that come in. Associate editors then choose specific reviewers or, in some cases, act as reviewers themselves. These days, reviewers are given standard forms to fill out regarding manuscript quality, and this helps make the final decision to accept or reject more straightforward. But the head editor nonetheless has the final say—the end responsibility always rests with him or her. The process is not democratic; at best, it is dilute plutocracy—imperfect, elitist, often inefficient, but effective.

Having the power to decide which papers are publishable means that an editor shapes both the scope and direction of a journal. Good editors are thus invaluable to a field and deserve far more recognition and reward than they get. The flip side to this, however, is that weak or overly autocratic editorial "regimes" can cause significant damage: even in the recent history of science, "quality control" has sometimes been an unfortunate euphemism for intellectual despotism. To curb such opportunities, and to prevent burnout as well, most editors are elected or appointed for brief periods, normally three to five years. In practice, a goodly number are re-elected and reappointed, because few people wish to take on the burdens (and calumny) involved. This, too, is a reason why good editors are worth their weight in platinum.

The Review Process: A Step-by-Step Outline

In practical terms, the review process in science serves to direct the revising of acceptable, written research. This means that, to editors and reviewers, all manuscripts are first drafts. As an author, you *must* be prepared to receive comment, criticism, and requests for revision.

When your manuscript is received, it is first logged in by an editorial assistant and possibly checked for completeness. If anything significant is missing—for example, you've included only one printed copy instead of the required two or three—it may be returned to you without even reaching the editor's desk. If the editor has to do this, you will have earned a black mark. Otherwise, your paper will proceed to take its place in the queue of submissions that await initial judgment.

The editor will then briefly survey the contents of the paper (title, abstract, headings, and illustrations, usually) and decide if it fits with the journal's scope. If it does, he or she next chooses who should review the paper or which associate editor in the relevant specialty should make this determination. Potential reviewers are then contacted, either by (e)mail or phone, and the manuscript is sent to them. Along with the paper, a reviewer will receive a letter stating a preferred deadline for returning the manuscript, a guide for making comments, and an evaluation sheet, often titled something like "Confidential Reviewer Report."

This report has a list of questions that must be answered (yes/no, or good/fair/poor). Typical questions include, Does the paper present original scientific content? Has the material appeared in any previous publication? Are there significant errors in fact, logic, or argument? Is the title accurate and sufficient? Are the illustrations appropriate? Is the reference list complete? Can any portion of the paper be shortened or omitted without loss of content? At the end of the report are four choices, one of which is to be checked: publish as submitted, publish with minor revision, publish with major revision, do not publish. The first of these is there largely to keep the number of choices even; it is almost *never* used. More likely, if your work is at all significant and relevant and you have followed the instructions to authors, one of the two middle choices will be yours. If not, you still have options.

At this stage, the reviewers return their material to the editor, who must then make a final decision. Either the editor, or a managing editor or publications director, will then send a formal reply to the lead author. This will include a letter noting the decision with the principle reasons for it, a set of recommended changes, a requested deadline for resubmitting the paper (if it has been tentatively accepted), and the original printout copies of the paper with reviewers' comments on them.

The ball is once again in your court. You must decide where things go from here. To do this will require that you sit down and go over in detail what the editor and reviewers have to say. If your paper has earned a "publish with minor revision," then by all means proceed ahead. If, on the other hand, you receive a "publish with major revision," you have several choices. You can make most or all suggested changes and resubmit. You can accommodate some changes, provide point-by-point reasons (in your cover letter) why you find other recommended alterations unnecessary or invalid, and resubmit. Or you can officially withdraw your paper and look for another journal. If your paper is rejected, you can submit it to another journal as is, or you can use the comments provided by the editor and reviewers to reshape your material and then send it elsewhere.

Once your revised paper has been accepted, it will be copyedited to accord with stylistic conventions of the journal. The text will then be set, made into galleys, and returned to you for final proofing. At this stage, you should make only the most necessary changes—meaning actual corrections—not updates of your data, new insertions, alterations to the structure of the article, or rewriting of entire paragraphs. This also goes for the illustrations when they arrive in final form (either with the text or separately). Despite the advent of digital publishing, changes at this point remain expensive. Too many will hold up publication of your paper or report.

Points and Pointers

Editors and reviewers are your first primary audience. If you can satisfy them, then the chances are good to guaranteed that your work will be reasonably suitable for, and will find, its larger, secondary audience—your (jealous) peers.

For this reason, whatever you do to make the editors' task easier will likely prejudice them in your favor and thus increase (but never secure) the chances of acceptance for your manuscript. Editors very much appreciate certain signs of consideration. These include receiving articles that (1) are potentially within the scope of the journal or publication, and (2) comply with the instructions to authors regarding style and format. Neither of these two considerations is optional, for any author. If you fail to meet them, your manuscript is doomed, and, worse, you'll go down in memory as someone who wasted the editor's time (as well as your own).

Fortunately, these necessary conditions go together. Even before sitting down to write, or at least during the earliest stages, you need to decide where you're going to send your manuscript. If possible, you'll want to have a first choice and a backup choice as well. Selecting a particular journal (or other specific outlet), meanwhile, is the *only* way you'll be able to comply with the second consideration above, since different journals nearly always have different specifications. This is a brutal fact of scientific life. It would be much better—for everyone involved, but especially writers—if this were not the case, if standards were imposed across the board for individual fields. Perhaps some day they will (though divine acts are rare). For now, however, take the time to find out the specifications for your particular journal or other publication and follow them. Note that such guides are usually provided for journals on the inside of every issue, and for other publications (e.g., symposia transactions, annual volumes, conference proceedings) by the editor or editorial staff directly. It helps to copy these, or print them out, and keep them handy to refer to as you write.

I would advise, very strongly, that you not go to war, or even enter the field of conflict, over stylistic details. Editors and reviewers, being indelibly flawed humans like many of us, are prone to certain pet peeves regarding specific points of usage. Where one will always change "since" to "because" or "while" to "whereas," another will delete every occurrence of "the fact that" or "in consideration of." Copyeditors, too, may be under orders to make alterations along these lines. Such changes are nearly always trivial—yes, unimportant. They derive from the great "schoolmarm" (Mencken) tradition in American letters and are based on bits of folklore about "proper usage." Except where they *truly* alter the meaning or make your prose flat and monotonous, you should simply accept (i.e., ignore) them and move on to more important things. The halls and stairwells of editorial offices across the land may well be stained red from battles over such positions, but it is blood needlessly spilled. Literary tics have been disobeyed by nearly every great writer (for wonderful examples, see the last chapter in Joseph Williams's book *Style*). You are better off without them and without worries over them.

Regarding the actual submission of manuscripts, meanwhile, it's essential to check out what each journal requires. In the entire history of the journal, reaching back to the 17th century, there has never been a single standard for preparing and submitting work, and there aren't any signs of such a miraculous phenomenon appearing anytime soon. Certainly, there are patterns, overlaps, similarities. But it will prove an unfortunate error to assume that the format, font, citation style, and so forth, either don't matter or are the same among different publishers. Likewise, simply sending in your paper and hoping the editors, blinded by the brilliance on display, will make whatever adjustments may be needed may not turn out the way you envision. The world of journals is complex, messy, demanding of your attention. As an editor once put it to me (long ago, let me say): "Be a grownup; follow directions."

One ugly fact for scientists: you can submit to only one journal at a time. Unlike in much of the humanities, simultaneous submissions remain *streng verboten* in science. From the writer's point of view, this is a profound disadvantage—imagine spending several months or more writing a paper on a time-sensitive subject, several months more having it reviewed, provisionally accepted with major revisions, then, after a year has passed, finally rejected. Your material is now dated and must be reshaped for another journal, which might have been very interested in the first place, but must now decline your paper in its original form, because a member of the competition has, during the interim, submitted an article on a similar topic . . . This type of situation does occur, from time to time, and can definitely leave scars. Fortunately, it remains the rare exception.

On the other side of the divide, the journal system in science would probably collapse overnight if it had to engage in overt competition for individual papers. To a degree, this is because of the need to safeguard the proprietary nature of research itself. The lack of simultaneous submission, that is, protects authors from having their work known and discussed by too many competing researchers. Obviously, this is no small consideration. The reality is that, as authors, scientists are both strengthened and weakened by the present system. This may well change, once the Internet becomes a prominent, perhaps the main, medium for publishing new science. Indeed, it is already changing for specific fields, as I will discuss later on (see chapter 15). But there is much heavy machinery and bureaucracy that needs to be moved. The journal remains sovereign, both in the print and online worlds, and will probably continue its reign for some time to come.

The Dignity of a Reply: Necessary Attitudes about Criticism

There is no doubt about it: dealing with the review process can be difficult, both emotionally and practically, but especially emotionally. Going over, point by point, what others have found inadequate in your written work may well be draining. It is especially hard for new or unseasoned authors, who have not yet developed an appropriately thick skin and whose work is perhaps more likely to earn a "publish with major revision" or "do not publish" reply. But it is likely to be the case for almost any author, at some level. Scientists are not known for the graces of courtesy and tact when commenting on the work of others. You may feel judged, embarrassed, even humiliated—then angry, wronged. On top of this is the added time and effort that will be required to modify your paper or report, another reality that can inspire frustration.

These are very natural responses, and we are all prey to them. They are part of the rites of passage for every scientist, and they are sometimes hard to bear. Yet there are certain ways to deal with them that can help minimize their negative impact and maximize the benefits and maturity you can derive from criticism in general.

First of all, consider this truth: reviewers and editors have not offered their comments about you, the person, but instead about an inanimate, inorganic object—the manuscript. This object is an entirely separate reality: it, not you, went through the mail and sat on a series of desks for evaluation. The manuscript came from your hands, certainly, but now it has a separate existence, and it is *this* existence to which all criticism is directed (a reason why the best criticism is always given in the third person). To the degree that you grasp this, and keep yourself off the page, you will be able to evaluate the comments you receive with reasonable distance and balance. This is an extremely valuable skill to have as an author.

Second, understanding the separate reality of the manuscript will help make it clear that commentary by reviewers and editors is intended to make the paper better, more fully acceptable to the client readership. Criticism is *not* aimed at a paper's destruction, but instead at making an improved piece of science. Granted, this is not how things are ordinarily expressed—usually, we must face the music of judgment: "The following aspects of your paper have been deemed unsatisfactory for the reasons given." Yet the goal is not to chastise, but instead to raise the level of what you have done to a more elevated stratum. Remember that the primary task of the editor is to ensure the quality of his or her journal, which is only as high as that of the articles contained within it. He or she therefore truly wants your paper to be as relevant, well-written, and complete as it can be (within certain constraints, of course), so that the light will shine that much more brightly on both of you.

You don't have to accept every suggested change to the manuscript. On the contrary, if certain criticisms appear invalid to you, let the editor know, by all means. You should send a separate letter explaining your response, point by point, with your resubmission, after making other changes you feel are relevant. In the rare cases (and they are, indeed, rare) where a reviewer may have overstepped the bounds of professionalism and allowed personal reasons (politics, vendettas, hobbyhorses, grinding axes, etc.) to dictate criticism, the editor is obligated to send the paper to another evaluator, and he or she may well strike the original reviewer from his or her list in the future. The reasons for this are, again, fairly straightforward. Maintaining quality requires a degree of vigilance in all aspects of the review process.

If your paper is rejected, do not despair—act. Find another journal. This is easier said than done, of course, since you'll have to reshape the paper in certain ways. But it is an essential response, both for you and your field. Editors frequently return the manuscript with a brief discussion of why it was rejected and, occasionally, what might be done to make it acceptable. Use this as helpful advice for submitting elsewhere. Alternatively, consider using portions of what you've written as the kernel for a different paper, appropriate for another journal. Cannibalizing what we've already done is a fully legitimate, and often practiced, form of expanding our options.

As a successful scientist, you are likely to be a professional author—writing is an essential part of your professional responsibility—and therefore you need to conduct yourself professionally in all situations. In all your correspondence with editors (you should never be put in touch, directly, with your reviewers), and with anyone else involved in the review and publishing process, it is *absolutely essential* that you keep your cool, remain courteous, and speak to the point. This will not only serve you in good stead in all your external dealings, but it will also help you achieve a certain useful distance from things. The authorial voice of professional calm is an extremely useful lie. No matter how heated a situation may become, it will impose dignity and help you stay in control. Using this voice is how you can best defend your work and uphold an image of command and competence. "Comments by this reviewer appear biased, unhinged, and irrelevant" is not the sort of response an editor will feel inclined to accept. Rather, couch what you say in tones that appeal to reason, that will make an editor feel he or she is being addressed as an intelligent, rational third party: "Comments by this reviewer, though well considered, are invalidated by the following points."

Being a professional means, above all else, communicating like one. Rage if you must, bestride private vales of smolder and fume, but stick to the high ground and cooler climes in your dealings with others. Moreover, on a wholly pragmatic level, keep a paper trail, that is, copies of everything that may pass between you and a publisher, so that if anything ever goes seriously wrong, you have the evidence on your side.

In the end, to complete the circle, perhaps the most practical advice of all rests in the arena of expectation. If you send in your paper knowing it will draw criticism and will have to be revised, you are definitely ahead in the game. This may not be easy. Secretly, at some level, we all hope to be that lucky (but apocryphal) exception whose article or report is immediately accepted, and accepted intact. The truth, however, is that nearly all articles submitted are first drafts to the review process. Therefore, plan for revisions—emotionally and temporally.

Finally, consider this fact: acceptance rates for major journals generally range from 8% to 60%. Periodicals like *Science* and *Nature* represent the lower range of this spectrum, as might be expected. Much higher percentages (>40%), however, are typical of the major specialized journals for individual fields and, especially, subfields. It is a good rule of thumb that the higher the prestige of a journal, the lower its acceptance rate. But what this means, exactly, will vary greatly among fields. Overall trends have shown a decline in acceptances by international journals that are widely read but also that the number of periodicals has grown each decade, as individual fields further specialize, as interdisciplinary directions develop, and also as top journals divide themselves into several offspring that focus on key subfields. In short, the world of the journal has become perhaps more dynamic than ever, in part due to the Internet. Informal surveys (not only my own), meanwhile, suggest that as much as 80% of all papers offered to publishers eventually find a home in print, in some form. Despair over rejection or required revision, however understandable in the short term, is not a practical long-term response.

✳

7. THROUGH A FLASK DARKLY: PLAGIARISM, FRAUD, AND THE ETHICS OF AUTHORSHIP

Why Does It Happen?

This is a brief and rather blunt chapter. It needs to be. The occurrence of unethical behavior in the lands of scientific authorship, especially forms of fraud and plagiarism, has continued to grow and broaden. While it remains a small part of scientific communication overall, and appears linked to historical realities in some measure, it is also an inarguable blight that must be unacceptable under any circumstance. This is because it qualifies as anti-science.

Thinking intently about fraud and plagiarism returns us to fundamental things. Though science is done within a great array of institutions and complex social structures, employing tens of millions of people and hundreds of billions of dollars, all joined by a gigantic web of scholarly, political, economic, and cultural priorities, it still has at its heart a single ethic: trust.[1] Scientists trust each other that the work they do, including the communication of their research, tries to add in some way to the corpus of knowl-

---

1. This point has been made many times, by many individuals. A highly relevant publication that begins by emphasizing the role of trust is the US National Academy of Sciences report *On Being a Scientist* (third edition, 2009). This can be downloaded from http://www.nap.edu/catalog/12192/on-being-a-scientist-a-guide-to-responsible-conduct-in.

edge. For any field, its knowledge is not merely a product but vital nourishment on every level. Trust is therefore both personal and collective, private and institutional. Today, moreover, it is global. The public shares in it, as do funding agencies, tenure committees, students, teachers, and a great many others. Stealing from the body of scientific knowledge while pretending to enhance it makes one, in literal ways, a parasite.

Every time a paper is written, an intellectual contract is signed in sweat (often with the miscible fluids of blood and tears). The contract says the writer has done this work and reported it accurately; whether the writer speaks untruly in private or about his or her age and weight is irrelevant. But submitting a paper or other communication to a publishing venue constitutes a public act, one performed in front of every other researcher in the discipline. If the submission is published, a second contract is signed, this time with the larger world. Breaking these contracts is a declaration of failure and a notification that one is not a scientist but an impersonator.

Why cheat, then? The question seems simple enough. Of course, it isn't. Some people cheat to outperform others. Some do it as a form of stealing. Some want a shortcut, to avoid putting in the full effort required. Some desire to appear better than they are, or better than they feel themselves to be. Others wish to create a result or finding when none was forthcoming from the actual data. Still others may be lured by the thrill of breaking taboos. And still others may feel the transgression to be trivial or, depending on their background, nonexistent. In every case, however, cheating is chosen as the best way to attain a type of success, *no matter what the risk.*

Does all of this apply to scientists? Absolutely. Why would intelligent, well-trained researchers, interested in advancing their prospects, commit plagiarism or fraud at the risk of permanently damaging, even obliterating, their career? The risk must somehow seem *worth it.* For those brought up and trained in Westernized systems of education, this calculus, which views plagiarism or fraud as a career move, is badly warped. It is warped for a number of reasons, one of which is that it damages science. It does this in two ways.

In a wholly practical way, when exposed, fraudulent or plagiarized work can weaken or invalidate other work based on its claims and findings. Some papers with falsified or plagiarized data, for example, haven't been found out and retracted for years, by which time dozens of other communications may have appeared utilizing such information (not all studies can be easily repeated to test data veracity; trust remains a factor here, too). There is a larger context as well. At a time when scientific knowledge is under suspicion and even attack by large parts of the public—over such areas as climate change, Darwinian evolution, genetic modification, animal research, among others—plagiarism and fraud provide sharp weapons to the opposition. In their work, scientists are assumed to exemplify a relentless integrity, a kind of moral high ground. This is why cases of misconduct so often make headlines; the failure is felt to be larger than that of an individual, a kind of exposé of a flawed system that claims to be the guardian of truth. Scientific work in some eyes becomes the product of just one more special interest.

The calculus that views plagiarism or fraud as a career move is warped for another reason. Clichés aside ("You are only cheating your*self*"), an act of plagiarism or fraud injures one's own discipline. It does this by damaging morale and wasting the time, and therefore resources, of editors and other scientists. It can also weaken public support, indeed even provide ammunition to the enemies of science as a whole, who can then point to corrupt and untrustworthy acts. Furthermore, it spatters the toxic muds of doubt and disrepute on one's own department, university, institute, and even, in some cases, home country. For countries where modern scientific work and publication are relatively new, the impact can harm all of these institutional levels at the same time. Such places are still in the process of trying to establish their credibility and advance their standing. Fair or unfair, cases of plagiarism or fraud can be a considerable setback, harming the integrity of the country's entire research system.

Finally, the chances of being found out are now far higher than they have ever been. The digital era, that is, may make cutting and pasting, fudging and fabricating, easier than ever, but it has also created a veritable library of plagiarism-detecting and comparison-making tools and services, created by companies that do this for profit and are therefore quite good at it and getting better all the time. When these services were first tested on particular journals, between 2005 and 2008, enough plagiarism was unearthed that, within just a few years, all major publishers of scientific material turned to them for help, and their journals now use them routinely.

Historical Factors

This discussion of the why of plagiarism and fraud brings us to the topic of historical reality. There are two parts of this reality to point out. The first has to do with the competitive side of science, which has grown more intense over the past several decades. There are many signs of this, even at the contextual level: decreasing success rates for grant applications, falling percentages of new PhDs with jobs at graduation, a plateauing in US government funding for research, among others.[2] We all know that scientific work can be a demanding and high-pressure profession, especially for younger researchers who are just starting out and therefore vulnerable. Finding a good postdoc position, a teaching or corporate job, the security of tenure, grant funding, good graduate students, and publication in a status journal—all of this is challenging at the least. Add to it the hope of achieving something important, making a real contribution, and gaining recognition for it, and you have a situation that tends to place rivalry eye to eye with collegiality.

Certain fields attract special attention, including media accolades (in the United States, this tends to be the biomedical realm). It can be both encouraging and dispiriting to witness weekly stories of the ubersuccessful, who presumably roll in research support by the barrelful and appear in the news sections of *Science, Nature*, or the *New Scientist*; in the pages of major newspapers; in books; on television; in TED lectures; and on film. Such luminaries bring the brilliance of attention, yet cast long shadows. They are heroes to their institutions, but objects of envy (and other emotions) as well. They also serve as troubled and troubling models for the more ambitious. Celebrity culture, that is, has come to exist in science, as in so many other domains of modern society, and can evoke dissatisfactions of varied type. Its most erosive effect, perhaps, is to make more ordinary researchers feel rather like protozoa.

The second part of the historical reality that needs to be noted has to do with the globalization of science and scientific publication. We are, in fact, since the late 1990s, living through a new era in the evolution of scientific work and thought. We may nod our heads at the mention of globalization in science, but it is a development of immense and unprecedented importance with direct implications for scientific communication. Consider: for the last three centuries, reaching from its beginnings in the 17th century to the late 1990s, modern science was ruled by a handful of Western and

2. See, for example, J. Weissmann, "The Ph.D Bust: America's Awful Market for Young Scientists—in 7 Charts," *Atlantic*, February 20, 2013, http://www.theatlantic.com/business /archive/2013/02/the-phd-bust-americas-awful-market-for-young-scientists-in-7-charts /273339/; and S. Rockey, "Comparing Success Rates, Award Rates, and Funding Rates," *Rock Talk* (blog), *NIH Extramural Nexus*, March 5, 2014, http://nexus.od.nih.gov/all/2014 /03/05/comparing-success-award-funding-rates/.

Westernized nations, confined mainly to Europe and North America, and later on including Russia and Japan. That era is now, as they say, history. Contemporary science takes place throughout many parts of the world, from Turkey to Brazil, China to Uganda, and continues to expand.

You probably see where I'm going with this. More than a few countries have not yet fully installed or do not yet fully enforce the ethical standards that are integral to modern scientific authorship and publication. Furthermore, the globalization of research is largely predicated on the use of English, now the acknowledged global language of science. Countries are in various stages of dealing with this situation. English-language teaching and learning are not what they need to be in many places, and competence in the language by scientists is also variable, though improving rapidly.

In short, history has helped provide conditions that make the ability to plagiarize very easy and the motives to do it appear almost rational. This is not at all the same thing as rendering such acts inevitable, let alone acceptable or lacking in responsibility. On the contrary—when someone is caught, the reputational impacts are now likely to be global, for the individual or the institution or even the country. Plagiarism and fraud are worse crimes today than ever before.

What Is Plagiarism?

Fraud and fabrication of data are fairly clear in terms of what they involve: the purposeful altering or invention of data, whether this be touching up a photograph or adding measurements that were never taken.

Plagiarism is something different. Defining it, in precise terms, can be challenging in the sciences. Why? Because it can mean stealing someone else's text, images, tables, data, experimental design, research procedures, methods, and ideas. On the simple side, the direct copying of material without any attribution or indication of source counts as robbery. This doesn't change if only a few words are deleted or replaced from the original. Altering the order of copied phrases or sentences doesn't at all reduce the fact of theft either. Nor does any other kind of rearrangement or slight adjustment. To coin a phrase: to copy another's words is to raid and rob his or her effort.

But there can be complexities. An inadvertent form of apparent plagiarism can occur when a citation is accidently erased or omitted, whether by the author or by a software problem. This does happen, though it's impossible to say how often. The only ultimate protection here is to be meticulous in rereading your draft before submitting it. Trying to defend against an accusation of plagiarism by claiming you are the victim of negative serendipity isn't likely to be persuasive. Another possible problem can emerge from the fact that, as writers, many of us absorb at some point—perhaps as a result of some recommendations and exercises in this book—memorable phrasing, even entire passages, while forgetting their original sources. If we're aware that the words are *recalled*, not created, we can make sufficient changes to prevent any allegations of theft.

Such potential difficulties call forth a question: How *much* of a text must be copied before it can be called plagiarized? What if only a handful of phrases are copied but scattered throughout the body of the new text? How *many* words have to be the same in order for an accusation of plagiarism to stick? And what about self-plagiarism? How grave is this?

None of these sorts of questions have a universal answer. Nor are there fixed standards even for single fields. Recommended guidelines certainly exist, but they are often quite general and sometimes vague. The Council of Science Editors, for example, in its *White Paper on Publication Ethics*, defines plagiarism as "piracy that involves the unauthorized use or close imitation of the language . . . and thoughts of others . . . without permission or acknowledgment."[3] Clearly, this won't help us a great deal; indeed, it adds yet another layer of uncertainty ("close imitation"?). Meanwhile, on the stricter side, I have seen guidelines that define plagiarism as beginning when six words in row are copied. This seems a bit extreme, given that phrases of six or more words can be quite generic and independently generated—for example, "quantitative approaches to improving data quality," or "ancient DNA from skeletons dated before 6,000 BCE," or even "plagiarism involves the verbatim copying of material without quotation marks or attribution." Extreme or not, however, this six-word standard tells us something: articles are now being microscopically scrutinized (by relevant software). Editors are looking for even the smallest signs of copied material.

So it's helpful to consider a few examples and how editors today might

3. Council of Science Editors, *CSE's White Paper on Promoting Integrity in Scientific Journal Publications, 2012 Update*, 3rd rev. ed. (Wheat Ridge, CO: Council of Science Editors, 2012), section 3.1.3, http://www.councilscienceeditors.org/resource-library/editorial-policies/white-paper-on-publication-ethics/.

view them. Here are two sentences from a published paper on global sustainability:

Human and natural systems interact in a multitude of ways. Quantifying the services that ecosystems provide for societal needs (such as clean water, nutrient cycling, and recreation) helps assign value to natural components for humans.[4]

Human and natural systems interact in a great many ways. Assigning values to ecosystem services that support societal needs (e.g., clean water, cycling of nutrients, recreation activities) is one way to generate data on natural components in a human system.

Does the second pair of sentences represent a plagiarized version of the first pair? Yes, it does. This shouldn't be hard to see, since only a few words have been changed and the original sentence structures are largely maintained, though not entirely. Let's look at another version:

It is typical for natural and human domains to be in complex, interactive contact. Can such interactions be quantified? The answer is yes if we are able to assign a rank or value to each service that a particular ecosystem offers for human benefit.

In this case, the writing is too different for the claim of textual plagiarism. Yet the content is largely the same. So piracy has actually taken place, right? Not necessarily. There isn't really enough evidence to say this. We need to look at what follows, to see if the original article was used as a detailed template, if the same facts, data, conclusions are used.

One final version is needed to complete the exercise:

Natural and human systems are interrelated at many levels, some of which are cooperative, others in conflict. For purposes of understanding which interactions may contribute toward sustainability, it helps to create a scheme able to identify ecosystem services that benefit society and to quantify their unique contribution.

4. J. Liu, H. Mooney, V. Hull, S. J. Davis, J. Gaskell, T. Hertel, J. Lubchenco, et al. "Systems Integration for Global Sustainability," *Science* 347, no. 6225 (2015): 965.

Comparing this pair of sentences with the first pair, we can see that there are differences in style, wording, and content. There are new details ("cooperative" and "in conflict") and new terminology ("sustainability") that, while seemingly small in overall volume, nonetheless add enough to forestall any accusation of plagiarism. Enough authorial work has been done here to create a new writing with new information. The author may choose to add a reference, but he or she doesn't have to do so.

This brief exercise is meant to show several things. First, it provides a reasonable example that answers an essential question for any writer: How can I tell when I'm paraphrasing or plagiarizing? Or, in another formulation, What have I really changed, and what have I left the same? Second, the samples above make it clear that plagiarism does not have to be word-for-word thievery; it also occurs when a majority of important terms are repeated. Third, and on the other hand, plagiarism does *not* always happen when a few sentences on the same topic seem to echo those in another paper. Likeness does not always mean larceny. Scientific authors sometimes use each other as a source of local inspiration. This should be fine—it can help one overcome writer's block—and where it applies to introductory or highly general material that anyone in the field might invoke, it may not even need a reference *if* it involves small volumes of text and doesn't become a *modus scribendi*, constantly relied on for a section in your paper. At that point, you are appropriating too much of someone else's writing.

But if your paraphrasing of another paper involves an interpretation, idea, or conclusion, then you *must* indicate the source, the author or authors who had it before you, offered it in trust, and from which you have benefited.

How Often and By Whom?

Fraud and plagiarism in science have been growing. The hard data are relatively new but convincing, while the anecdotal evidence is immense and depressing. Editors and publishers have spoken of a "retraction epidemic," particularly since the late 1990s. They are certainly the ones to know. For those of us who work outside the editorial mines, it can be sobering (and sometimes entertaining) to read through the blog *Retraction Watch* for a list of recent examples (my own favorite: a study on generating natural gas from animal waste, published in the journal *Clean*, whose lead author had to re-create data because his notebook was "blown into a manure pit"). But if we may sometimes doubt the stories, we must still face the numbers.

Early test runs employing antiplagiarism software, conducted by such publishers as Elsevier, Taylor and Francis, and the Nature Publishing Group, revealed that editors were right, indeed more right than they would have preferred. Rates of plagiarism varied in these tests from a low range of about 2%–5% of articles in a specific sample to as much as 15%, with levels reaching over 20% in a few cases.[5] These figures are for papers submitted to prestigious journals by scientists from a range of nations (data were not divided according to institutional affiliation or country). But percentages given in some studies of selected non-Western countries (e.g., China) have been as high as 31%.[6]

It's important to point out that the software used for these determinations can and does yield false positives. It is not a replacement for hominid evaluations. The practice now followed by editors is to let the software flag articles with apparent overlap and then have these examined by the human eye for final judgment.[7] This is both sensible and necessary. But it also adds to the burdens that editors and their staffs already carry. Thus it wastes the time and effort of these crucial gatekeepers, on whom (let us put it this way) much good will depends.

There have also been serious studies about rescinded scientific papers. One such effort, published in 2012, searched through more than 25 million publications archived by the PubMed database (biomedicine, mostly), finding 2,047 retractions. Moreover, a close look did reveal some interesting, if disturbing, results. Over 67% of retracted papers were withdrawn because of fraud or suspected fraud (43%), duplicate publication (14%), and plagiarism (10%). These are largely forms of intentional misconduct, which soared in number between 2001 and 2011. They tended to concentrate in top journals, with high impact factors, for example, *Science, Nature,* the *New England Journal of Medicine,* but were spread among many others as well. Geographic distribution? Lead authors came from 56 nations, with the United States accounting for nearly half of all cases involving fraud or

5.   D. Butler, "Journals Step Up Plagiarism Policing," *Nature* 466, no. 7303 (2010): 167.

6.   Y. Zhang, "Chinese Journal Finds 31% of Submissions Plagiarized," *Nature* 467, no. 7312 (2010): 153.

7.   Y. Li, "Text-Based Plagiarism in Scientific Publishing: Issues, Developments and Education," *Science and Engineering Ethics* 19, no. 3 (2013): 1241–1254.

suspected fraud (Germany and Japan were next), China taking the lead for duplicate publication (followed by the United States and India), and the United States, China, and India leading for plagiarism (Italy and Turkey were next). The overriding conclusion? Scientists from developing countries are *not* the main perpetrators. Misconduct is still a form of expertise most highly developed in the West.[8]

Now, it might strike you that 2,047 retractions out of 25 million documents is pretty paltry evidence to confirm a problem. Quantitatively, you would be correct. Realistically, however, you would not. These retractions represent only the ones that were found. Undoubtedly, given the youth of the relevant detection net, a good many slipped through. In more than a few cases, moreover, it took years, even a decade, to discover the problem. Thus many articles citing fraudulent papers were directly affected. Not only has a mere fraction of the total transgression been identified, but many of the bad papers both discovered and undiscovered have created entire genealogies of damage—think of an impact on a sheet of glass, whose fractures keep proliferating outward. This damage also includes large quantities of wasted time, effort, money, and faith for those who must monitor the literature.

Yet there is a bit of a bright note here. Online plagiarism software can actually be used by anyone to check papers before they are submitted. This means that scientists and students can help take the lead on the problem by checking their own work, thus preventing false positives. It has been suggested, as well, that university departments do this as a matter of course for all papers submitted by their faculty, a policy not everyone is likely to feel warm and happy about. It does appear quite possible, however, that journals will begin requesting such self-checks before too long, as a kind of virus protection, once the software is more advanced and universally available. Obviously, this would be far from foolproof. But the fools would then have an extra layer of deception to pursue, a fact that might give some second thoughts and that, in any case, would make offenders doubly guilty.

8.   F. C. Fang, R. G. Steen, and A. Casadevall, "Misconduct Accounts for the Majority of Retracted Scientific Publications," *Proceedings of the National Academy of Sciences* 110, no. 3 (2012): 17028–17033.

A Necessary Conclusion

These statements take up only a few of the major issues surrounding fraud and plagiarism and their possible causes. They are enough, however, to point you in a particular direction, toward a well-defined and hard-edged principle: scientific authors today must turn away from all forms of plagiarism, no matter how small or seemingly incidental and no matter what the temptation. Scientists do not often use quotation marks, so the necessity is to rewrite something and provide it with the anchor of a reference. But to borrow is to steal. I say this for all the ethical reasons mentioned above, which need to be understood, but also for the very real and practical truth that acts of plagiarism by competent scientists can bring concrete damage to an otherwise worthy career.

I say this *especially* for scientists with English as a foreign language (EFL). Here, the impact of being caught can reverberate well beyond one's own career, affecting the standing of many others in one's home nation. Detection databases now have tens of millions of articles from thousands of journals to check any paper against. Authors found guilty of plagiarism are duly recorded; their names go on a list that can be shared with other scientific publishers. I'm sure I don't have to explain what this might mean. Having future work put under the lens of suspicion and doubt, even (since we are still dealing with human beings) stained with a drop of negative bias, seems a high price to pay for a few slips of the pen or keyboard. Once caught, forever known.

It would be foolish to pretend that only minds of venal substance are tempted toward fraud and theft. Many an excellent scientist, including some I have personally known, has felt the urge to nudge his or her data or to adopt another's admirable words. Resisting such attractions today, especially those of plagiarism, in the new digital era, counts not only as loyalty to science and the virtue of trust, but qualifies as a form of self-protection. Using the work of others as a model is one thing; embezzling it is something quite different.

※

Part 2. Communicating Professionally

WHERE, WHAT, AND HOW

8. PROFESSIONAL SCIENTIFIC COMMUNICATION: WHERE DOES IT HAPPEN?

Contexts of Communicating

Every scientist should know something about the greater context in which he or she works and communicates. This context—the institutional setting of research in society—is dynamic and, over time, has expanded greatly. Even today, we might be forgiven for thinking that research mainly lives within the halls and towers of academe. This, however, is to mistake a branch for the tree.

In rich, developed nations, scientific inquiry actually occurs in four main settings: industry, academia, government, and private nonprofit institutions, plus consortia among these. Please note I am not talking only about funding for research—in both wealthy and developing nations, most such funding comes from government (at least 60% of it in the United States)—but about *where* research work is actually done. One's research setting is the stronger determining factor with regard to the forms of communication that are demanded or expected.

Let us take these four settings in order of scale. We begin with industry. Despite a long-term overall decline in research support by private firms (spending, personnel, facilities, other forms of investment), industry still ranks above all other settings in terms of the total research it does. The industry setting includes companies ranging from giant pharmaceutical firms working on dozens of new drugs to the start-up focused on a single new catalyst for biofuel production. Over time, there has been a big shift in research work from the largest companies to smaller, specialized firms. The older model of an IBM or Merk or Exxon with its own large-scale research center, employing legions of PhDs working at the outer edges of frontier knowledge, no longer holds. Big companies haven't given up scientific inquiry, not at all. But they are not as invested in it as they once were. They now depend on the findings from—or the acquisition of—small, nimble firms able to focus more efficiently and intensively on key research problems. In brief, the ecology of industry research is complex, dynamic, and evolving.

Academe is the second-largest source of research in the developed world. This also comprises a diverse category of institutions, varying from large research universities, both public and private, down to small liberal arts colleges and, increasingly, community colleges as well. The bulk of investigative work continues to take place in research universities, particularly those with major graduate programs. Besides the labs in campus buildings, where professors and postdocs carry out experiments, build models, test ideas, and train apprentices, scientific inquiry also happens in intercollegiate research facilities, college-affiliated medical centers, corporate labs, and, in certain fields like ecology and geology, distant parts of the globe. Today, there are many opportunities for scientists to do part or all of their work at other universities that offer joint projects, or at special institutions in their field, like the Large Hadron Collider (Switzerland), various research stations in Antarctica, and so on. They may also found their own private companies and work simultaneously in the two worlds of industry and academia. Indeed, some universities even encourage this (though things can get fairly muddy and messy when it comes to intellectual property). It is still true that academic research gets most of its funding from government, in the form of grants. But there can also be strong support in some fields from industry, private foundations, and wealthy donors.

Next, we come to government labs and research centers. These tend to perform a much smaller part of a country's total scientific inquiry than do industry or academia. But this doesn't at all mean they are less important. Government research has been responsible for some less-than-trivial innovations—the Internet, lasers, satellites, GPS, for starters—as well as a few discoveries, like those related to nuclear energy, elements 93 to 118, or the nature of the lunar and Martian surfaces (plus a few hundred thousand details about the rest of the solar system). In developed countries, government supports research for a number of reasons: social benefit, national defense, international standing and status. Social benefit, to be sure, is a broad goal. Agriculture, aerospace, energy, medicine, the environment, coastal and geological surveys, materials science, and a number of other domains may all be pursued in individual departments. The choice and scale of such research depends directly on the priorities of the relevant government, which, of course, means political decisions. In the United States, for example, there has been a long-term emphasis on biomedical research; the budget of the National Institutes of Health (NIH) is more than three times that of any other government agency, with the Department of Defense in second place (I am not the first to point out a certain irony here) (Celeste, Griswold, and Straf 2014, 2–3).[1] In Japan, on the other hand, priority research areas include the physical sciences and engineering, while China puts an emphasis on these same areas, plus mathematics.[2]

Finally, we arrive at a fourth category. This is private, nonprofit research work, encompassing three main types of institutions: independent research institutes, major foundations, and NGOs (nongovernmental organizations). These settings tend to have an activist character and to carry out inquiry in specific fields rather than in a full range of disciplines as universities do. Their intent, baldly but honestly put, is to improve the state of the world in a particular area, for example, by reducing a nasty disease, finding ways to increase biodiversity in a specific ecosystem, or improving stewardship of Southern California coastal areas.

Research institutes are sometimes called universities without students. They often employ postdocs, as well as established researchers, and commonly have a tenure system. Their facilities are usually excellent and may share land or space with a university, in which case they may be required to help train students. Though endowed (founded by a pot of money from someone), they compete for government and other grants to enhance their capabilities.

Foundations, meanwhile, are a different animal. There are two basic types: corporate and noncorporate, both of which give money to programs and institutions, rarely to individuals, scientists or otherwise. Both

1.   R. F. Celeste, A. Griswold, and M. L. Straf, eds., *Furthering America's Research Enterprise* (Washington, DC: National Research Council, National Academies Press, 2014), 2–3.

2.   UNESCO, *UNESCO Science Report 2010: The Current Status of Science around the World* (Paris: UNESCO Publishing, 2010), 9.

types provide grants rather than seek them, but may have a number of re-searchers on staff to help design and operate the different focus areas for giving and to judge grant applications. Corporate foundations tend to offer grants or other assistance in areas related to their own business, but they may also support the arts, education, or community development. Giving by corporate foundations whose parent company is involved in science- or engineering-related work might extend in several diverse directions—for example, the ExxonMobil Foundation has the focus areas of malaria pre-vention and treatment, math and science education, and economic oppor-tunity for women. The General Motors Foundation, as another instance, is dedicated to education, health and human services, environment and energy, and community development. Noncorporate giving by private foundations has grown enormously since the 1990s, and in many cases, these foundations have extended their goals well beyond the home coun-try into the international and global realms. Not strictly research-oriented, these foundations are interested in social betterment and so deal with pol-icy issues that bring science together with real-world problems. A particu-larly large number of such foundations concerned with research are con-centrated in the areas of health and education. This is especially true in the United States.

NGOs are true issue-oriented, activist organizations. Their number has grown enormously in recent decades, and this includes NGOs that do scientific work. Young and midcareer researchers who wish to make a dif-ference in a way other than publishing papers in big-time journals have not been hard to find for such work. NGO research has become concentrated on topics identified as having major importance for the future: food and agriculture, water availability, biodiversity, conservation of wild lands, disease prevention, and maternal and infant mortality. Young researchers with a strong social conscience are drawn to such topics.

NGOs in all such areas have scientists on staff. If large and well-funded, like the World Wildlife Fund, Nature Conservancy, or Save the Children, they operate throughout the globe, in many countries, with assets in the billions of dollars and budgets well into the hundreds of millions of dollars. This allows them to have many scientists working for them, who collabo-rate with academics, government researchers, local experts, and others. Because of their focus, NGOs survive (and sometimes prosper) mainly on support from memberships and fundraising. They are beholden to those who share their vision for improving the world. Indeed, the overall con-text of advocacy can pose a challenge for researchers, who must obey two masters: the cause of their organization and the need to keep research from being politicized.

Globalizing Science: The Importance of Developing Nations

Again, all of the above relates to wealthy nations. Since the early 21st century, however, a growing portion of it has also come to be true for what are called emerging nations. These are countries that have industrialized and built modern economies to a significant extent and whose research institutions have been partly modeled on those in the West. Such countries include China, India, Brazil, Mexico, Turkey, and Indonesia. In these countries, it is common for research work to occur in industry, academia, and government, as well as in research institutes. The relative importance of each setting varies a good deal, however; in some nations, research is less advanced or less prestigious in universities than in industry and government. One factor that *does* carry across most of these countries and that distinguishes them from wealthy nations is a high degree of state control over the direction of scientific work. Unlike the United States, Europe, Japan, and South Korea, where scientific institutions are autonomous, emerging nations tend to have state-centered systems where the government uses a variety of policy tools (directed funding, taxation, oversight, accreditation, etc.) to support those areas and programs it views as priorities. These priorities tend to favor practical innovation over pure research. China, for example, increased its support for R&D more than fivefold between 2000 and 2010, yet the percentage devoted to pure research fell continuously while that targeting development programs and projects grew greatly.[3]

At the same time, there have been growing demands in emerging nations for researchers to build an international presence. This means publishing papers in international journals, attending and presenting at important conferences around the world, forming multinational collaborations, and participating in major international research projects. Researchers in these countries now frequently collaborate with their counterparts in the United States and Europe, and they are also involved in large-scale, frontier efforts, such as ITER (International Thermonuclear Experimental Reac-

---

3.   Y. Sun and C. Cao, "Demystifying Central Government R&D Spending in China," *Science* 345, no. 6200 (2014): 1006–1008.

tor), the European Southern Observatory, and the Large Hadron Collider, among others. Thus, the communicational demands for researchers in emerging nations have expanded correspondingly. And indeed, if we look at publication counts, we see these have risen an enormous amount for developing nations as a whole, from around 17% of all science and engineering articles in 2000 to something like 45% by 2014.[4]

For less developed nations, the situation is different. Here, the industrial sector has not yet grown to where it can support, or justify supporting, scientific research as an integral part of business. In countries like Cambodia, Tajikistan, and Bolivia, overall investment in science and technology is quite low, less than 0.4% of GDP overall, compared with emerging nations at an average of twice this.[5] Research tends to be dominated by the government and to be focused in only a few areas that are immediately relevant to the country's development needs (e.g., agronomy). Yet there is growing collaboration between scientists in these countries and those in both emerging and wealthy nations.

The globalization of science has progressed to the point that scientists today, whether they begin in the United States, Brazil, or Tanzania, may well journey through an array of research settings during the length of their career. They may therefore need to become adept at communicating in new and perhaps unfamiliar ways, both regarding peers and the public. I will go over some of these ways below.

What All of This Means for Communication

Industry

Many scientists in industry do a lot of writing, as much or even more than academic researchers. What they write, however, doesn't often appear in journals. There are exceptions to this, of course. Periodicals like the *Journal of Pharmaceutical Technology and Drug Research* and *Minerals and Metallurgical Processing* are places where industry scientists inevitably make a

---

4. See, for example, National Science Board, "Academic Research and Development," chap. 5 in *Science and Engineering Indicators 2016* (Arlington, VA: National Science Board, US National Science Foundation, January 2016), http://www.nsf.gov/statistics/2016/nsb 20161/#/.

5. See, for example, National Science Board, "Research and Development: National Trends and International Comparisons," chap. 4 in *Science and Engineering Indicators 2016*.

regular appearance. The ones who do therefore understand the particular requirements of the scientific paper (as discussed in this book). They also advance their long-term status and career possibilities. Yet most industry communications by researchers fulfill internal company needs. Such communications come in a variety of formats: memos (brief or long), reports and presentations to management, internal newsletters, research summaries, press releases, patent applications, drug approval applications, documents for stockholder meetings, website material (much variety here, with pretty graphics), and a great deal more.

Depending on the specific industry and a company's status or market share, there may also be communications aimed at informing, swaying, calming, or enthusing the public, even policy makers in government. A good example: several major energy companies who use their websites to offer sophisticated analyses of global energy trends and forecasts for free in downloadable format. BP (British Petroleum) has been publishing its annual *Statistical Review of World Energy* since 1965. Not only is this document now an important, highly anticipated reference for analysts and officials around the globe, but BP has added to it a number of other informational booklets, energy-charting tools, and energy fact sheets on regions and certain countries, as well as transcripts, podcasts, and videos of related speeches by company luminaries—all available for free.

In short, firms now use the Internet in a number of informational ways to connect with the larger world. Long gone are the days when company websites were mainly ads or commercials. Businesses of every scale are now trying to be good "informational citizens" by offering knowledge and advice. Like the rest of scientific communication, this domain will continue to evolve, to become more sophisticated and diverse and therefore demanding of high-level writing and presenting skills.

Academia

More research may take place in the corporate sector, but academic scientists generate the great majority of publications for journals, conference proceedings, books, websites, blogs, and other venues that, together, largely define the communal voice of science. While they have come to dominate the communicational dimension to science, university researchers don't often view themselves as carrying the literary torch. Nonetheless, it is mainly in their hand.

Academic researchers don't entirely have a choice in this. They live and work within a system of rewards and recognitions that continues to count publications as primary career capital. We know it as the publish-or-perish system, which extends from grant application to journal paper. Each field has its own hierarchy of journals, and a single paper placed in the pages of a top periodical is generally worth several that appear further down the ladder of status. But the status system is not necessarily the last word. A paper's worth and fame can depend on how significant the work actually is to the field; not all truly important research is published in the upper-stratosphere journals, even though citation counts suggest that it is. Academic research today, as I have already noted, is global and continues to globalize further with each passing year. It is simply not possible for a few dozen (or even a hundred) journals to capture the essence and scope of all truly significant work. Researchers therefore need to keep their eyes and minds open to the larger field of publication in their discipline and how it may be evolving.

The role of communication extends much further than research publication itself. Academic scientists have responsibilities that reach well beyond the written dimension alone. They must also train each new generation of researchers; they are the first parents of science's future. Thus they need to be good or excellent communicators in the classroom, face-to-face. If mediocre or poor at this, they create damage, however temporary. If they are excellent, they do the opposite. They can refuel the fire in those with the drive to deeply learn about the physical-biological world and enhance human knowledge about it.

The responsibilities of academic scientists don't end here. To be truly successful research scientists in the 21st century, they must do still more. They must present their work at conferences, in lectures and talks, and, when it makes sense to do so (and it does more and more often), in communications with nonscientific audiences—the public, the media, primary and secondary schools, museums, investors, donors, and (the final rub) elected officials.

Compared with a few decades ago, researchers have an expanding variety of ways to fulfill such communication. The tried-and-true traditional methods of lecturing and speaking directly to public audiences remain more important than ever. New visuals are expected, and needed, to fill out the "inform *and* entertain" dimension, particularly digital slides and images, video and audio, brief animations. Top scientific communicators still weave a certain "wow" factor into their talks and lectures, even if in sub-dued fashion. The era of the TED talk demands as much.

Yet the key to reaching large audiences in the digital age lies else-where, with social media. Science blogs have been quite successful with many parts of the public, but they are not for every researcher, as they re-quire time and thought and can become a second career. This can lead to many positives, but it tends to work best for those who are already good-to-excellent (not to say efficient) communicators, thus a limited subset of the total *corpus scientia*. Much less demanding in this way, yet capable of reaching even larger audiences, is Twitter, whose messages are confined to a mere text-bite of 140 characters, thus a brief announcement, observation, comment, or other informational message. I will have more to say about this option, and others as well, in chapter 15.

Suffice it to say that the days when a university scientist was expected to deal only occasionally, if at all, with the outside world, beyond his or her office, lab, and immediate colleagues (those who speak the same arcane language), are now over. For a variety of reasons, not all of them worthy of revelry, science is in the public eye now more than ever before—an eye that can be skeptical or doubting. While the reasons for this lie beyond the scope of the present volume, they create a context in which communicat-ing with the public often places the researcher (like it or not) in the role of representative for science as a whole.

All of the above means that, as communicators, academic scientists live and work in a new world compared with their forebears. It is a fact that comes with its own opportunities, risks, and responsibilities. While re-searchers can still carve out a career in a confined social space, as in the past, this is fast becoming a niche existence. The truly successful academic scientist now has an expanded communicational world, one defined by an ability to speak and write in different types of discourse—professional, semiprofessional, informal, nonscientific—to both live and virtual audi-ences.

Government

If you work as a government researcher, chances are you will be called on to write and speak in various capacities. One factor will uniquely apply to what you produce: it is in the public domain, which means it can be freely used by anyone, anywhere, anytime, as he or she sees fit. In this sense, government publications are the opposite of proprietary work by the corporate/business sector. Their task is to make scientific information widely available. At the same time, this doesn't mean they translate everything into ordinary language for universal consumption. On the contrary, most research is written professionally, in professional discourse, and is therefore intended for professional scientists.

Depending on the field, this may include reports (both short and long) and parts thereof, as well as articles in government journals, special papers, conference proceedings, book chapters, and more. In short, government researchers can be involved in writing no less diverse than is common for successful academics. Indeed, these researchers often publish in academic journals and attend academic conferences too. They are active members of their respective fields. Another unique aspect to their work is that, in many cases, writing material for publication in government documents is part of their job—they are, so to speak, guaranteed the opportunity to be authors. The degree to which this is true can vary greatly, to be sure. Yet it is hardly rare that researchers who have worked in government service find later employment in academia.

Government scientists are now asked to also author materials intended for general audiences. Indeed, this task has grown since the early 21st century to be a sizeable part of researchers' work. There is a strong demand, at least in advanced nations, that federal departments have an Internet presence, usually in the form of websites, often with numerous levels of information. If your field has attracted public interest (positive or negative), you may well need to help assemble explanatory guides, fact sheets, timelines, materials for teaching, or other kinds of outreach materials. A good example is the information put out by the US Geological Survey on its "Earthquake Hazards Program" web page. This page offers separate information on the most recent quakes worldwide, as well as seismicity maps, hazard maps, archives on past earthquakes, educational topics, a "For Kids" section, educational slides in some cases, and a great deal else. For a major quake with significant damage to human settlements, the survey will prepare a special poster (in PDF format) that summarizes key data, provides essential maps, and discusses the event in language that is supposed to be fairly accessible. Here is an example:

> The January 3, 2016 M 6.7 earthquake near Imphal, India occurred as the result of strike slip faulting in the complex plate boundary region

between India and the Eurasia plate in southeast Asia. Focal mechanisms for the event indicate slip occurred on either a right-lateral fault plane dipping moderately to the east-northeast, or on a left-lateral fault dipping steeply to the south-southeast. In the region of the earthquake, the India plate is moving towards the north-northeast with respect to Eurasia at a velocity of approximately 48 mm/yr.[6]

Unfortunately, there is a problem. Despite all the good intentions on display, as well as the excellent information offered, there is no real entrance for the general public. The language here seems aimed at a mythical region between a trained geoscientist and a well-educated layperson with a good science background. The problem, of course, is that this Middle Earth counts as fiction; only the geoscientist can comprehend the paragraph. For the public, a more accessible alternative would run something like this:

The January 3, 2016, earthquake near Imphal, India, was a major event of magnitude 6.7. This is large enough to cause significant damage in those areas with maximum shaking, given the unreinforced nature of most buildings. The quake was the result of active faulting along the complex boundary between India and Eurasia. This boundary marks the ongoing collision between the Indian plate, which continues to move northward at about 48 mm/yr (4.8 m/1000 yrs, or 480 km in 10 million yrs) and the Eurasian plate.

Yes, I have both added and removed some material. The former gives some crucial information for general readers, telling them why this might be considered a major event in human terms. The translation of 48 mm/yr into larger terms brings this figure first into daily experience and then into a geologically meaningful context showing that huge amounts of movement take place on this scale. The deleted part, about fault details, can be added later in the discussion. The first few paragraphs in a writing like this, however, should be for nonspecialists. If it is public information, it needs to be for the public.

The larger point is that government researchers are like researchers everywhere. Few of them are trained to write for general audiences, and

6.   US Geological Survey, "M6.7 Eastern India Earthquake of 03 January 2016," accessed April 25, 2016, http://earthquake.usgs.gov/earthquakes/eqarchives/poster/2016/20160103 .pdf.

doing so doesn't merely involve an adjustment, like shifting down a gear or adding a diluent. Here is not the place to say what is needed (chapter 19 is the place). What should be stressed, instead, is that government scientific agencies are more in the business of public communication than ever, and scientists who work for them need to be aware of this for the possible sake of their own careers. Informing the public is now taken quite seriously, if still selectively. This is partly because it addresses directly the ever-sharpened sword of accountability that hangs over federal and state researchers.

As a result, government scientists will often need to learn how to write for both professional and general audiences. In this, they will be in a similar position to their academic counterparts, with the difference being (again) that it is in the job description. The challenge for scientists in both settings is that, while they may well be at least functional in scientific expression, they are unlikely to have had much training in science writing and speaking—that is, translating their professional knowledge into common forms of discourse. As I will discuss in chapter 19, this requires altogether a different skill set, a very valuable one to own.

Research Institutes, Foundations, and NGOs

This final category, hodgepodge though it is, largely combines the elements described above for academic and government researchers. Research institute scientists focus mainly on seeking grants and publishing in professional journals. Those who work for foundations, meanwhile, are typically on the other side of the grant barrier—reviewing rather than seeking. Foundation scientists can be involved in much more than this, though. Various kinds of outreach communications are also needed, especially to media contacts, government officials, and the public, as well as to grantees and other foundations. Foundations are very much in the business of self-branding. Though they are grantors of funding, and therefore stimulants to competition, they are also in rivalry with one another to a certain degree. It may be better to give than to receive, but, from this perspective, it is even better to give and receive influence. The same things can largely be said for NGOs. Their work is no less involved in outreach, and sometimes more, since one of their chief goals is to raise awareness (not just money) about a specific problem area, issue, or injustice.

Location, Location...

Wherever you decide to ply your scientific field today, there are likely to be diverse demands when it comes to communication, both writing and speaking. This is challenging, to be sure. But it also brings a host of new opportunities compared with only a few decades ago, when the public presence of the average researcher was equivalent to that of deep-ocean fauna. This chapter charts the major communicational obligations that now exist, though the scope of these will vary among disciplines.

The chapters that follow tend to focus on forms that relate most immediately to the academic sector. This is justified by the reality that the journal paper continues to be the most fundamental type of contribution by researchers, including those in all the sectors we have discussed. Knowing what goes into such a paper, and being able to produce one, also means that you have the communicational skills needed to conceive, organize, plan, and write any high-level technical document, most of which are adaptations or expansions of this basic form. Other chapters, devoted to presentations, online work, scientific translation, and communicating with the public, apply to all sectors where research and its representation take place.

The fundamental point is that successful career scientists have more work to do in this area than in the past, and they can reap real benefits from it. The digital era gives us the ability to make contact with hundreds, thousands, perchance even larger numbers, of potentially interested readers and listeners, not only in our own country but around the world. Whether viewed from the perspective of the 1870s or the 1970s, this counts as an astounding change, breathtaking at the very least. Again, as I have made clear, nearly every location for research work offers such opportunity. Scientists just starting out, as well as those underway, might take note: the world may be getting smaller, but digital realities also make it much bigger in the audiences and contacts that can be created.

*

9. THE SCIENTIFIC PAPER:
A REALISTIC VIEW AND
PRACTICAL ADVICE

Yesterday and Today

The modern scientific article, though the core of technical communication today, began life discreetly, as the lecture and the letter. These were transcribed in the earliest journals, the *Journal des Sçavans* in Paris and the *Transactions of the Royal Society of London*, both of which appeared in the 1660s as formal outlets for presentations given before each respective society, the Académie des Sciences and the Royal Society of London. What is striking about these early articles, to the modern eye, is their splendid variety. It is a wonder to see what teeming heterogeneity is there, a medley of both of subject and approach, from laboratory research to speculative thought experiments, from microscopic observations to field reports on distant lands. The science of 300 years ago required such diversity to be expressed. It contained in vitro what would eventually grow into the most prolific enterprise of knowledge production the world has ever seen.

To such variety, the scientific paper has ever remained true. No single definition can encompass its full reality. "A written report describing original, replicable research" is fine as a high-altitude description, but nothing more. The two critical terms here—"original" and "research"—change considerably in definition across disciplines. Recall that there are many tens of thousands of refereed scientific journals in current publication, and their number grows by as much as 8%–9% a year.[1] A bit of variety might therefore be expected. In my own field, geoscience, everything from mathematical simulations to revisions of previous fieldwork counts as research and has its place in the literature. Other disciplines recognize neither of these species as acceptable or relevant, but instead focus on lab work. Science is no more a single method to reveal truth than is art.

Thus, it is *essential* for every scientist to explore the literature of his or her own field. This is really the only way you can gain a *realistic* idea of what counts as publishable research and what doesn't. It means learning what constitutes the "primary literature" and the "secondary literature," and what might straddle the increasingly porous boundary between. The world of scientific publication is both conservative and malleable and is now in a process of significant flux. Investigating the literature is therefore an important part of your research.

Types of Papers

Scientific publication is actually a vast, evolving cosmos today, one that matches the breadth and diversity of technical effort itself. This, of course, is just as it should be. Many different types of "papers" now exist in the world of scientific publishing, some results-oriented, some not. Such variety also reflects the fact that many journals (actually, editors) like to remain flexible in terms of what they might offer their readers. In practical terms, meanwhile, this means that there are a number of different kinds of articles wherein a scientist may discuss work in progress or express opinion about the work of others. The following list gives just a few of the more common article types.

Results-oriented papers frequently include

- *Research Articles.* These are the mainstay of technical publication, actual reports of new work intended to introduce new knowledge in a specific field. Depending on subject and focus, these articles can vary from a few

---

1. See J. G. Paradis and M. L. Zimmerman, *The MIT Guide to Science and Engineering Communication* (Cambridge, MA: MIT Press, 1997), 178; and L. Bornmann and R. Mutz, "Growth Rates of Modern Science: A Bibliometric Analysis Based on the Number of Publications and Cited References," *Journal of the Association for Information Science and Technology* 66, no. 11 (November 2015): 2215–2222.

pages to 20 pages or more. Those of greater length may be referred to in a journal as "reports," "articles," "original papers," or some other term.

- *Letters/Short Communications.* These are brief research papers (<5–6 printed pages) that report new work, but require less space. Such articles can also carry material that is usually peripheral to the main thrust of a journal, but deemed of direct interest to readers. Other terms used for this type of article are "research notes," "brief communications," and "short papers."
- *Commentary/Forums.* This category can include different types of papers, often brief debate-oriented commentaries or discussions. They may expand on earlier articles, seek to clarify or amend specific points, or offer criticism based on new work by the author(s). Some journals routinely provide space for limited, formal debate and will include both a critique and a reply. Other common titles for this type of writing are "discussion/reply," "opinion," "viewpoint," and "debate."

Types of journal articles that are less results-oriented may include

- *Letters to the Editor/Correspondence.* Most premier journals carry a section with this title, containing technical responses to earlier published material. Such letters usually serve the purpose of debate and criticism. Less often, but increasingly in some quarters, they respond to an editorial, book review, or other results-free writing.
- *Review Articles.* This type of report presents a critical survey or overview of recent research and thinking in a particular field. It is aimed specifically at keeping scientists up to date with findings in their own and related fields. Such articles are commonly longer than a research report and contain a large reference list. However, it has become fairly common for journals to publish "minireviews" on topics of high current interest. Review articles tend to be commissioned by journal editors.
- *Book Reviews.* Many premier journals carry a book review section, plus a list of "books received." Such reviews are usually quite short and follow a style and format specified by the journal editors.
- *Editorials.* A small but increasing number of journals publish opinion pieces, usually one page long, on topics that might range from recent discoveries or ethical concerns to federal budget cuts and public controversies. In the past, editorial writing was the exclusive province of editors; however, it is a growing trend to encourage scientists to contribute to this forum.

As I say, these are among the most prevalent types of articles in scientific journals. But they hardly define the entire field. There is significant variety well beyond the types listed above. Moreover, journals are themselves highly diverse in what they offer: some regularly provide examples of every article type just mentioned, while others publish only research papers. Most lie somewhere in between.

In all cases, therefore, it is essential for you to survey the journals in your field to see what, exactly, the range of your options is regarding publication. All the examples given above are put through a review process of some sort. They can all be found in bibliographies of research papers and should be listed on any curriculum vitae. In short, they usually count as part of the "primary literature," which I take up in the next section.

Primary, Secondary, and Other Literature

Traditionally speaking, the "primary literature" constitutes the first publication of research results, commonly in a well-recognized journal. "Secondary" refers to any subsequent appearance in print of such results, in such venues as review articles, conference proceedings, book chapters, and so on. Primary publications are peer reviewed and counted directly toward tenure or research posts, both inside and outside of academe. The secondary literature, in the past, was usually unreviewed, ephemeral, and commanded far less prestige. It was viewed as much less important to career advancement.

Over the past several decades, these distinctions have begun to soften. This has happened in direct response to the changing, indeed broadening, realm of scientific publication as a whole. Primary results now appear not only in premier journals but also in proceedings of important one-time symposia, special volumes on designated topics, monographs, government reports, book chapters, and more. New journals, too, are being launched continually. The position of these is usually uncertain for the first several years, but might be high from the beginning if they are the first to cover an emerging subfield.

The secondary literature in science, meanwhile, has come to include a huge array of publishing opportunities that, again, vary considerably from one field to another. Abstracts, transactions, conference proceedings, local bulletins, posters, newsletters, websites, and other such outlets normally count in this category. Yet, with the rapid shift of science communication

to the Internet, a host of other types of publications have come into existence, as described in chapter 8. In recent years, these outlets have become
more important, as government, industry, NGOs, and other nonacademic
centers of research have sought both professional and public audiences.

Article Structure: Parts and Their Purposes

It bears repeating that there is no standardized template for scientific
papers applicable to all journals and all fields (nor should there be). To convince yourself of this, go to your nearest university library and browse
through periodicals from different disciplines. You'll find that while some
require a fairly set article structure, many others do nothing of the sort.

Most journals, however, require that papers contain some version of the
following parts: title, abstract, introduction, background, methods/mate-
rials, discussion of results, summary/conclusion, references, acknowledgments, and supplemental materials. What follows is a brief review of each
of these parts. In reading through this, and in your perusal of the literature,
remember that each major section in a paper centers on a distinct type of
content and, to a significant degree, uses a different style of writing.

Title

The title is the most important phrase or sentence (except for all the
others). The most-read portion of any paper, it announces the article, telling other scientists whether they need to read the paper. Try to keep it
short; use key words that will help your paper be indexed properly (otherwise, it may be lost to its potential readership). Titles are usually phrases.
In some publications (e.g., the premier journal *Cell*), sentences and questions are acceptable. Read through several issues of the publication to
which you intend to submit for models to emulate (if necessary). If you're
having trouble choosing a title, try jotting down one or two working versions during the writing of the paper, then revise at the end. Here are some
examples of good titles:

LABRYRINTHINE PATTERN FORMATION IN MAGNETIC FLUIDS
(short and attractive)

THE SOURCE AND FATE OF MASSIVE CARBON INPUT DURING THERMAL MAXIMA (dramatic and to the point)

A ROLE FOR PROTEIN PHOSPHATASES IN LONG-TERM DEPRESSION OF THE HIPPOCAMPUS (well-phrased)

LABORATORY MODEL FOR DEEP EARTH CONVECTION: HOW IMPORTANT IS A THERMALLY HETEROGENEOUS MANTLE? (states both the topic and the "problem")

Poor titles, meanwhile, might look like this:

TO CREATE A PROTEIN-BASED ELEMENT OF INHERITANCE (Who or what is doing the creating?)

VOLTAGE-DEPENDENT LIPID MOBILITY FACTORS IN THE OUTER HAIR CELL PLASMA MEMBRANE (Noun phrases are too long; try VOLTAGE-DEPENDENT MOBILITY OF LIPIDS IN THE PLASMA MEMBRANE OF OUTER HAIR CELLS.)

AN UNNATURAL BIOPOLYMER (Too little information.)

A SUSTAINABLE ROUTE TO THE CREATION OF MICROCELLULAR MATERIALS USING CARBON DIOXIDE TO PRODUCE FREE-STANDING GELS WITH LOW BULK DENSITY AND NANOMETER-SIZE CELLS (Too much information; save everything after "materials" for the abstract.)

Abstract

The abstract is the second most read portion of any paper—and, increasingly throughout science, a crucial publication in its own right. Indeed, abstracts are doubtless *the* most widely exchanged and distributed form of professional scientific writing in the world today. They are often the only published evidence of conference talks, presentations, and research updates. They are frequently excerpted and republished in reference volumes. In many online journals, they appear in a box when the cursor moves over an article title. They are now included in most online bibliographic databases, a major new aid to research. And abstracts are also forms of knowledge "capital" that scientists trade among themselves almost as readily as they do greetings (or criticisms). All of which highlights the considerable importance of this unique, condensed form of written communication.

Write your abstract last, after the rest of the paper is already drafted. If, as the saying goes, brevity be the soul of wit, then the abstract requires clever chiseling. A good abstract is more than an executive summary or a series of generalizing statements. It is much closer to a minipaper, a compressed version of an article or talk (which, in a sense, justifies its separate publication), minus figures and tables. Think of the abstract, therefore, not as an add on but instead as a stand-alone, an entity that, if decapitated from the rest of the paper, would convey its bodily substance. In many cases, after all, it will be all that a reader sees.

This may sound formidable. It needn't be so. Writing good abstracts doesn't depend on gifts from above, but on observation (of good examples) and practice here below. Try to follow the basic order of points in your article. In many cases, one to three topic sentences for each section of your paper are sufficient. Be sure to include scope and importance of topic, basic approach used, some specific data, and most important conclusion(s). Keep abbreviations to an absolute minimum. Don't include too much hard data (it clots the narrative)—select only the data that help establish the "problem" or support the main conclusion, perhaps just enough to tickle the readers' interest so that they might go on and read or search out your full paper.

Perhaps the most common problem in creating abstracts is the urge (doubtless felt by all) to cram everything in and get it over with—an impulse that, when allowed its day, will lead to long, heavily burdened sentences that need to be read several times, before true confusion sets in. For example,

> Analysis of historical and recently available 2-D reflection seismic data along the eastern and northwestern margins of the San Juan Basin reveals a close relationship of Laramide-age basement involved faulting with fracture orientation in the vicinity of several fractured Mancos Shale reservoirs used as a basis for an important recent structural study of oil occurrence in the basin.

It makes good sense to begin an abstract with a short sentence—a brief and clear statement, in fairly *general* terms, giving the importance of the subject or focus of the study:

> Seismic data reveal a close relationship between basement faulting and the orientation of fractures responsible for oil production in parts of the San Juan Basin.

Or

> We present results of a seismic data study confirming a close relationship between Laramide-age basement faulting in the San Juan Basin and oil production from fractured reservoirs.

Abstracts need to be just as readable as a well-written article. Here too, do unto your readers as you would have them do unto you. Think about what information is absolutely necessary, and what isn't:

> Biologists now believe that the first colonization of land by eukaryotes resulted from symbiosis between a photosynthesizing organism (phototroph) and a fungus.

A nice opening sentence. But "phototroph" doesn't appear anywhere else in the abstract. It isn't needed and, in fact, adds a touch of confusion, because it signals the reader to look for further mention of this term. Delete it and the sentence is excellent. Here's another example:

> For some time now, there has been considerable experimental and theoretical effort aimed at understanding the role of normal-state phases in high-temperature cuprate superconductors.

Again, a reasonably good opening. But the abstract is no place for "conversation." Try, instead,

> Considerable experimental and theoretical effort has been aimed at understanding the role . . .

In *all* cases—this is very important—check issues of your targeted journal for examples of what is accepted. Study the instructions to authors, as they might have information that pertains specifically to abstracts (e.g., length and format).

Finally, be aware of an unwritten rule followed by some editors: namely, that the abstract should not repeat verbatim any sentences in the main text. I seriously doubt the average editor has the time or inclination to check this in detail for every paper, but he or she may be tipped in the direction of doing so if, for example, the first sentence of your abstract is identical to that of your introduction, which follows immediately after. Moving a few phrases around should be sufficient to take care of this.

Introduction

Much more than a rhetorical welcome, the introduction is an essential
entry hall into the house of your paper. It needs to offer the reader a hos-
pitable and substantive reception. This means stating the problem or topic
you are writing about, why it is important, how you have approached it (in
general terms), and what is new about what you have done. A good introduction thus points up the gap in existing knowledge that the paper will
help fill. In most cases, it also provides a degree of essential background,
for example, an outline of existing knowledge on a topic, definitions of
terms, or a brief review of the existing literature.

Good introductions begin with a brief sentence that launches the topic.
This is immediately followed by supporting details:

> The development of chemotherapeutic agents for the treatment of HIV-1
> infection has focused primary on two viral enzymes: reverse transcrip-
> tase and protease. Regimens including agents directed at each of these
> biochemical targets are effective in reducing viral load and morbidity
> and therefore mortality. However, the long-lived nature of the infection
> and the genetic plasticity of the virus have made it apparent that new an-
> tiretroviral agents are required to deal with the appearance and spread
> of resistance. To address this issue, it may be important to consider the
> process by which viral DNA achieves insertion into the host cell genome,
> namely integration, which is catalyzed by HIV-1 integrase.[2]

> Subduction of the Juan de Fuca and Gorda plates has presented earth sci-
> entists with a dilemma. Despite compelling evidence of active plate con-
> vergence, subduction on the Cascadia zone has often been viewed as a
> relatively benign tectonic process. There is no deep oceanic trench off
> the coast; there is no extensive Benioff-Wadati seismicity zone; and most

---

2.   D. J. Hazuda, P. Felock, M. Witmer, A. Wolfe, K. Stillmock, J. A. Grobler, A. Espeseth,
et al., "Inhibitors of Strand Transfer That Prevent Integration and Inhibit HIV-1 Replication
in Cells," *Science* 287, no. 5453 (2000): 646. I have slightly altered the opening to this article.
In the original, the second two sentences are combined into one: "Although regimens including agents . . . , the long-lived nature of the infection . . ."

puzzling of all, there have not been any historic low-angle thrust earthquakes between the continental and subducted plates.[3]

A common way to close an introduction is with a description of what you did, plus one or two major conclusions that resulted. For example,

> This paper suggests a new approach to identifying inhibitors capable of preventing HIV-1 integrase catalysis. . . . In particular, diketo acid inhibitors were found to manifest significant anti-viral activity as a consequence of their effect on integration.

> In this report, we extend the study of previous authors by systematically comparing trench bathymetry and shallow seismicity for a worldwide sampling of subduction zones. Such comparisons lend support to the possibility of great subduction earthquakes off the Pacific Northwest coast.

Background

A background section is frequently needed to bring readers fully up to speed for the material that follows. Depending on the journal or publication venue you've chosen, this may or may not be a required part of your article. The background portion can take a number of different specific forms. In some cases, it is where a "review of the existing literature" or "previous investigations" should go. In other instances, for example, when you performed fieldwork, a section that describes your setting might go here. Depending on the specific subject, many disciplines require a theoretical section to help outline major assumptions, previous thinking, current hypotheses, mathematical bases, and the like. The only template to follow is what the publication requires. If this is unspecified, decide whether you actually need a separate section or whether you can roll the relevant information into the introduction.

---

3. T. H. Heaton and S. H. Hartzell, "Source Characteristics of Hypothetical Subduction Earthquakes in the Northwestern United States," *Bulletin of the Seismological Society of America* 76, no. 3 (1986): 675.

Methods/Materials

A methods section is required for studies that involved some type of laboratory work, whether based on experiment or measurement or both. A number of different titles are regularly given to this section, depending on the field: Materials and Methods, Experimental Methods, Research Procedure, Apparatus and Procedure, among others. All of these titles, however, are used for a separate portion of a paper or report that offers a description of the tools and techniques you used to solve the "problem" stated in the introduction. As such, it defines an essential part of many papers and reports for several reasons. First, it allows the reader to follow and, if desirable, repeat what was done—thereby permitting verification or questioning of the stated results, a fundamental tenet for all scientific research. Second, it permits an evaluation of how skillfully the work was designed and carried out, thereby offering a possible model for other researchers. Third, it places the work in a certain historical context—stating that these were the tools and the techniques that were available at this point in the history of the field—and therefore leaves the work open to fertile restudy when newer methods come along (as they inevitably do).

For these reasons, writing in this section must be especially straightforward, even cookbook, in style. This section requires a specific type of discourse not found in other portions of the article (and thus points up how complex a document the scientific paper really is). As always, consult your models of good writing. Evaluate how they handled their particular subject; gauge whether you might be able to repeat their experiments.

Discussion of Results

In the discussion of results, you say what you found and what it means. By this time in the article, readers will be prepared for your specific contribution. They will have been briefed on the topic addressed, what gap in knowledge it will fill, and how previous work has handled or overlooked it. They will also have taken a short tour of your laboratory or field setting and have been shown how you chose to go about your investigation. Now it is time to return to the lectern, present your data, and reveal its significance.

For laboratory studies, and sometimes for other types of investigations as well, this part of the report is frequently divided into two sections, results and discussion. In this case, your data should be presented and briefly discussed in the first section, with your major interpretations reserved for the second.

For nonlaboratory work, these two sections are sometimes used, but it is also common (in some fields, far more common) for authors to craft individual headings specific to their topic. In this case, findings and interpretations are often merged under each section. For example, individual data areas or subtopics (e.g., seismic information, gravity data, magnetic data, geologic structure) may have their own sections and subsections, each of which might include the presentation and interpretation of the relevant information.

If this is the approach in your field, examine closely how results similar to your own have been put down effectively by other authors. Pay attention to the use of illustrations and how these are able to collect and summarize information. Most often, you will probably have some idea of the types of tables, charts, graphs, maps, and images you might want to use. But a glance through your models or the literature in general might reveal specific forms that you overlooked and that suit your topic very well, or, alternatively, that suggest how you might creatively adapt a given form to your case. If you are using a results-and-discussion structure, beware of the impulse to dump information onto the page. Too often, inexperienced writers want to rush through the data section in order to get to the interpretations (the exciting part). But data are never just "data"—numbers, measurements, quantity. They have meaning, obvious or not. Any display of information will contain or suggest interpretations. These first-order interpretations are really part of your results. Here are some examples:

> The three tests performed as part of the initial experiment revealed an improved correlation between the root mean square of factor Y and the observed trend Z.

> Six of the RNA functional groups were found to make significant energetic contributions to the formation of the signal recognition particle complex.

> It is clear from Figure 3 that stresses are rotated up to 45° in the lower crust, due to deep detachment faulting beneath the Basin and Range province.

Any of these would do well in a results-type section. What then of the discussion, or its equivalent? Here, you would speak more generally:

> Results derived from the series of experiments reported here suggest [that previous correlations involving factor Y have underestimated the importance of . . .] or [the following conclusions: (1) . . . , (2) . . . , (3) . . .]

> These studies on the structure of signal recognition particle in *E. coli* provide a new understanding of conserved ribonucleoprotein elements. [Follow with specific points.]

> As shown here, both analytical and finite-element models of extension in the Basin and Range support the conclusion that rheological differences between the upper and lower crust determine fault geometry at depth.

Once you've put down your major conclusions, take a look back. Make sure your tables and graphics are appropriate and that they aim in the direction you want them to, that is, toward your final interpretations. Keep in mind the experimental nature of the writing process. Be open to the possibility of discovering new interpretations after you have graphed and presented the data and had a chance to look at it anew, in these forms.

Finally, the discussion of your results in a good place to express confidence. Talk about the larger implications of your work, whether these relate to theory, methods, or analyses. Don't be afraid to make a claim for the importance of what you've done—after all, you've worked hard and deserve notice. But don't overdo it either. Einstein never claimed to be "Einstein" (read his papers and you'll see). Watson and Crick stated only that the subject of their DNA paper "had considerable biological interest." Therefore, gauge the import of your work relative to that of the subject and what others have said. The two "infinities," self-negation and grandiosity, are the Scylla and Charybdis of scientific writers.

Summary or Conclusions

In a summary or conclusion section, you have an opportunity for bringing closure. Every article, whatever its story, requires an ending. In science, this usually means several paragraphs that briefly summarize what was done, what was found, and what significance the work may have for present and future research. This final section should not repeat data or go over detailed interpretations given earlier. Instead, it should link your work to larger ideas and issues.

A good concluding section provides balance to the introduction and a sense of change. This is true in two ways. First, in your final paragraphs, you should return to the topic with which your paper began in the first several lines and state what new knowledge or new thinking has been added. In other words, the topic remains, just as important as before (perhaps more so), but illuminated or expanded or refocused in a particular way. Second, recall that your introduction began generally and progressed to more specific points—your conclusion section should do this in mirror fashion, ending with the broadest statements.

Here's an example showing both these points, from the first and last lines of a published article:

> The composition of a planetary surface is an important indicator of its early evolution and subsequent chemical alteration. The surface composition of the jovian moon, Europa, for example, has been modified in a number of ways over time, including through intense bombardment by jovian magnetosphere particles . . . that could have effected change through radiolysis. The relative importance of this chemical alteration process has not yet been established for Europa.

> The abundance of H_2O_2, and the existence of an Na and O_2 atmosphere thought to be produced by energetic-particle bombardment of the surface, demonstrate that surface chemistry on Europa is dominated by radiolysis. . . . Predictions, characterization, and identification of surface chemical species on this planet must therefore consider radiolysis effects more closely than in the past.[4]

An effective conclusion tells the reader that something has changed, that science has been increased in some way.

4. R. W. Carlson, M. S. Anderson, R. E. Johnson, W. D. Smythe, A. R. Hendrix, C. A. Barth, L. A. Soderblom, et al., "Hydrogen Peroxide on the Surface of Europa," *Science* 283, no. 5410 (1999): 2062–2063. The quoted text has been slightly altered from the original.

Acknowledgments, References, Appendixes

Acknowledgments and references are required ingredients for any paper. They are the places where you most directly reveal the social reality of your work—where you show that you are part of a scholarly community and that you have depended on the kindness of grant-giving strangers (or, in some cases, friends).

When writing the acknowledgments and reference list, consult directly your targeted journal for style guidelines. There are no final standards in this area: different journals typically employ different styles. Thus, referring to the journal is the *only* way you can be sure you are doing things correctly. Acknowledgments offer gratitude to both granting institutions and to a possible range of individuals. If you're uncertain who to thank, look at examples from published papers to see who might be included.

The reference list is an essential element to get right. Most instructions to authors specifically note how the editors want you to write this part of the article. Follow their guidelines closely, to the last semicolon and period: this is a mark of being professional and considerate (careless authors force copyeditors to spend much time, money, and emotion in this area). If no stylistic guidelines are provided, refer to published papers in the targeted journal as models. It is a good idea to do this in any case to get an idea of how long your list should be and what types of sources are generally used. Some journals—not very many—allow for explanatory notes to be included in the reference section: once again, consult other articles to see how this is done.

Appendixes are occasional additions to papers. They are usually reserved for detailed offerings of data (especially in tabular form), mathematical derivations, discussions of innovative methods, or other aspects that are important to the work being published, yet would take up too much space (overwhelm or distract the reader) if included in the main portion of the article. An appendix is thus an opportunity, reserved for very special cases, to present more information than commonly occurs in the brief space allotted a journal paper. Some journals allow subsections within an appendix; others require that you divide these into separate appendixes or shorten everything as much as possible. If you have any questions about whether you should include an appendix, check with several experienced colleagues or the editor of the journal.

Supplementary Materials

Some journals allow or require you to submit other matter, often called supplementary materials. Thus far, it is much more common for a journal to give the author the option of providing such material, rather than making it mandatory, but as inclusion of it becomes more frequent and thus the norm, the result may be peer pressure adding to peer review for it to be made standard. Such is what I have heard from editors. But I refrain from predicting future events.

Be that as it may, there is much variety in the kinds of material that journals consider acceptable. Sometimes supplementary material includes a fuller set of data tables and explanation. Other periodicals may ask for more, such as a detailed description of your methods, additional text that fully explains a conclusion, more complete captions for figures, and so forth. Still others, depending on the field, will leave the doors open to an array of enhancements, from video to extra biographical information, although there may be limits to the size of files. As always, it's essential to read the instructions for authors.

Adding this option to the scientific paper is not intended to tempt you into making your work as author a greater burden and time sink. Instead, its aim is to expand the scientific value of your contribution. In a sense, it provides a way for you to add back some of the good work you had to leave out or cut from your paper. And there is one more advantage to consider: such material, in an era of increased worry about forms of misconduct, helps answer the concern for more accountability.

Outlines Can Help

Because the scientific article is usually firmly structured, outlines can help in the actual writing process. There are many approaches to building outlines—I've touched on some of them previously (see chapter 4, under "Organization")—but one of the simplest and most direct methods is to begin by writing down, in provisional form, the major headings for your various sections, for example, introduction, methods, sections for each data area or subtopic, conclusions. Or, if you can't decide on these yet, even tentatively, use the main parts of the article: introduction, background, discussion of results, conclusions. List the topics or major points you want to include under each of these, in any order you want, as ideas come to you. Then see if you can find a logical order for them, within each section. Sometimes—often, really—this will be enough to get you started writing. As always, if progress lags, consult your models, look through other papers in your targeted journal. It very often happens that papers and reports can be organized as hybrids based on other documents. Don't be shy to emulate or, selectively, even copy what others have successfully used: after all, they inevitably did the same. Above all, leave any outline open to change: it is a guide, not a template.

A cautionary note is in order. Many scientists (and nearly all manuals on scientific writing) make the error of assuming that results and discussion define the true core of any report. This is like saying that the torso is the most important part of the body. Without the other parts, it is meaningless and dead. All sections of a competent article are necessary and require the same amount of care and attention; there is no specific hierarchy among them. A careless introduction or inept conclusion is the writer's equivalent to bad lab procedure or misleading data. As scientists, we are taught that data and interpretations are our main business, certainly. But when we present our work to others, we become writers and speakers, and communication has its own demands.

Citations in a Scientific Paper: Their Meaning and Use

Citing other authors in a paper does several things. First, it offers accountability. It tells the reader that you are familiar with the most recent, significant literature in your area and that this literature has aided you in your work. Second, citation is a way to outline a community of like investigators—a collegium, if you will. Third, citations are a tool by which you express various degrees of agreement and disagreement toward the work of others within this community: colleagues can be cited favorably ("the excellent work of Barnes et al. 1987"), unfavorably ("Delpy [1994] failed to consider"), flatly ("has been the subject of numerous studies, e.g., Batts 1978; Resin et al. 1983; Foresby 1985, 1992"), and in qualified fashion ("the work of Jensen et al. [1998] requires further support"). Most documents employ several of these types—they are how scientist-authors rank their cohorts and competitors and position themselves toward them. Fourth, citation is also a way for making certain claims to originality, or, perhaps inadvertently, the very opposite.

These are complexities that flow from one another and that we all recognize in science, usually on an intuitive level. I make them explicit here to show, again, how multidimensional a scientific document really is, but also to help provide some awareness of the reasons for the way scientists write as they do. Here, too, that is, you can gain an added degree of control by understanding what is being done. Let's take several examples.

Case 1. Suppose, in the interest of being thorough and precise as a good scientist should, you are abundant with your citations, adding a reference to support every generalization, making sure you've included all relevant sources within the past 5 or 10 years, no matter how obscure. Perhaps you've just completed your PhD and are writing it into article form: you have all these worthwhile references that you've worked hard to compile and properly place—why not use them? Isn't this part of being complete, conscientious, and professional? The answer, I'm afraid, is no. Think of your references as a type of data: to be (or not to be) professional means being selective, offering your readers only as much as they need to reconstruct your work without any sidetracks or distractions. Piling up citations makes you look hesitant, conciliatory, and worse, derivative. On the practical side, it also renders your document varicose, clotting it up, making it physically difficult (thus inconsiderate) to read.

Case 2. Suppose, on the other hand, that you decide to include only a very few citations, to only the most well-known literature: what then? Another backfire. This type of referencing (or lack thereof) will suggest tones of intellectual plagiarism. The desire to appear foremost and original will make you seem guilty of fraud, a risk no good scientist should be willing to take. If your work truly is pathbreaking or exploratory, make this known in how you present your results and how you discuss the work of others. The height of unprofessionalism is to pretend that you have few or no peers.

Case 3. Let us propose, finally, the circumstance where an author cites his or her own work fairly often, or perhaps very often, as well as that of other colleagues: what blessings or sins arise here? In fact, self-citation is one of those unavoidable gray areas in scientific writing that can't be easily dictated either way. Referencing your own work too often (say, more than 20% of the time or so), or stacking your reference list with your own articles (e.g., more than 5 or 6 out of 30–40 total), will produce an image of unprofessionalism for the same reasons just given above. If you are one

of the only researchers in a particular area, however, you may be forced to
grant yourself a high degree of recognition. Even here, a degree of humility
defines the better part of valor: at all times, try to strike a balance between
your own contribution and your debt to others.

Coauthors: How to Organize and Manage
Contributions from Others

Research today is collaborative, and most scientific articles have more than
one author. Indeed, it is now common for papers to have at least three coauthors, and some include as many as a dozen (the record, in fact, is over
100). This should make us stop for a moment and think about what "authorship" means in this context—for it is clearly different than in other areas of intellectual work (imagine a literary essay with seven authors . . .). In
nonscientific fields, having your name on an article indicates that you did
most of the research and, above all, the actual writing: the idea of "the author" is still tied directly to the crafting of language, the putting of words to
paper. Not necessarily so in science. Authorship here is much more loosely
and variably defined to include all those who have made an important contribution to the work being represented, whether this be in the design and
execution of experiments, the overview and mentoring of work, or the actual writing. The masthead of any individual scientific paper is therefore
closer to a dramatis personae or list of participants.

This brings with it a series of decisions that are not always easy to make.
Friendship, collegiality, laboratory or departmental politics, institutional
traditions: all of these can play a role in determining who is to be included
and in what order. Generally, the first name given is recognized today as
the senior author, with progressively decreasing contributions assumed for
the order of names thereafter. This is sensible, yet it may not be true. After the first several names, how can you decide who goes where if 10 more
"authors" are involved? Diplomacy and negotiation—or allegiance to al-
phabetical order—are often required. Authorship, like research, is not a democracy, but it shouldn't be a dictatorship either.

Realistically, there are no final rules in this area—but a few guidelines
can help. Fairness is a good start (and finish): include as authors only those
who have actively participated in the intellectual work being represented.
This means those who generated real substance, who could accurately be
said to have conceived and worked through a certain portion of the re-

search and writing and who might therefore be able to defend it. It may help to define, precisely, what "authorship" means for *your* paper—what range of contributions, what degree of involvement in the research, what level of participation in the actual writing. Above all, to thy own collaborators be true: if a (prestigious) colleague contributes an idea or two at a seminar or during a hallway chat, does this merit his or her inclusion alongside those who have worked hard for months in the lab or field? Probably not, even if the idea proved pivotal in helping direct or focus the research. Instead, use your acknowledgments section to recognize this type of contribution.

At some point in your career, however, you may have to deal with authorial hitchhikers (dare we say parasites?). This is a lamentable, even deplorable, reality, but it is one that flows from the reward system in science, where publications count as career capital. It is still an occasional practice for the names of lab directors, department heads, graduate advisors, or other dignitaries to be added as a matter of course to any paper that emerges from within their jurisdiction, whether they had any substantive input or not. Another example is where a certain lab may routinely include the names of technicians on all papers. If and when you encounter such a situation, it is necessary, for your own career, to carefully evaluate the situation. Every article is born out of a series of social situations (this is very much part of science, too), and it is advisable to be a savvy traveler and participant in these. Pick your battles carefully: live to write another day.

Choosing a Journal

How do you decide where to send your work? *International Bulletin of Ornithology* or, instead, *Ozark Birdwatcher's Newsletter*? Prestige and wide distribution are frequent considerations. You'd like your research to make an impact, to reach as large an audience as possible. But it sometimes pays to be realistic, too.

Experienced scientists, whether in academia or a nonacademic sector, will know the literature well and likely have a journal or two in mind at the outset. If you are an early-career author, however, you may need to consider this question more thoroughly. Certainly you need to pose it at the very beginning of the writing process, since different journals have different stylistic specifications. In truth, if you're reasonably familiar with the literature in your field, a good part of the decision will be made for you:

such aspects as the subject, approach, and scope of your work will tend to narrow your choice down to a few publication possibilities. Beyond this, look for as close a match as you can between the nature and content of your research and what has appeared in these publications. Which journal might be most welcoming of your research (properly written, of course)? Where have colleagues working in the same field, on similar projects, published their results? Where have articles appeared that you might consider using as models or guides for your own? What about any new journals— including electronic (online) journals—that might be appropriate to consider? Such questions might help focus or finalize your search.

The nature of scientific publishing requires, however, that you choose one, and only one, journal at a time. Again, each publication has its own blueprint for manuscript style and preparation, and science does not allow for simultaneous submissions (i.e., sending your paper to several journals at once). If you've been rejected by one journal and are considering submitting to a second choice—after appropriate revisions, of course—make sure that you modify your paper to accord with that journal's specifications. You can also look at a rejection as a new opportunity: think about breaking the original paper up into smaller pieces and publishing them separately, for example, as notes or letters that comment on articles that have appeared previously. This is a fully legitimate and well-recognized outlet for research. It can also help lay the ground for resubmitting your original paper (again, in modified form) elsewhere.

Submitting the Manuscript (and a Final Statement)

For a great majority of journals, the online version is now primary, the more important edition compared with the print copy (if one even exists). This means that submitting a paper for publication is also mainly done in digital form. But not always and not entirely in every case.

Many journals still want you to submit one or more (usually two or three) hard copies of your paper. These are intended for reviewers, some of whom prefer to work with actual paper. Do not suppose that such individuals are doomed to extinction in the near future; they are not all over 70, by any means. Meanwhile, some periodicals will let you submit data tables, photographs, and figures in final, high-resolution printouts, which they will then scan. But this is far from the rule, so please don't assume it will be done. I say "please," because it is not pleasant for an editor to send back an excellent paper because stated directions for submission weren't obeyed. As I've discussed previously, editors will not accept your paper simply because it is high in quality and exciting in subject. They probably *want* to, but they also know it is a dangerous precedent to set in light of their future survival.

As things now stand, there are no universal standards for sending your work to the journal of your (or your lab director's) choice. One or more persons on your research team should therefore be assigned to learn the submission requirements in detail and to review the final paper according to them. More generally, it is part of the responsibility of every author to at least be aware of such requirements. These, in fact, may change over time, so the responsibility involves keeping up to date too. The Internet will continue to evolve, and with it, many aspects of scientific communication. We can certainly hope that things will become simpler and more efficient. We can't, however, take it for granted that this will be the case.

The reason we can't is that, along with the Internet and science itself, the journal article will continue to evolve. Already, as it has shrunk in overall size, it has gained new ingredients, as the brief discussion above on supplemental materials shows. Some papers, as mentioned, now appear online with readers' comments, which can add a number of different elements (be it debate or new data or references). In a word, the scientific paper has never stood still, historically speaking. Now that it has a new medium, it may well find new ways to grow and become even more useful.

✳

10. OTHER TYPES OF WRITING:
REVIEW ARTICLES, BOOK REVIEWS,
DEBATE/CRITIQUE

> The only thing worse or better than a bad or good review
> is the person who wrote it.—WALTER MATTHAU

Advice and Consent

Research papers are the core of scientific publication, to be sure, but the entire fruit is much larger than this. Most scientists know as much from their weekly reading experience. Premier journals today, and certainly the big international periodicals like *Nature* and *Science*, often carry a range of writing: review articles, letters to the editor, editorials, book reviews, meeting summaries, debate, obituaries, news of the profession, and more, in addition to research reports. Increasingly, in fact, journals have broadened the array of writing within their borders, or given new importance to types (e.g., the review article) that formerly appeared only occasionally. Such variety exists, in part, because of increased pressure to keep periodicals attractive and useful. Call it a policy of noble survival. Editors have consciously decided to make their publications responsive to the evolving communicational realities and needs of science—in essence, to make the journal itself even more a marketplace of ideas, a nexus of professional expression, delivery, and exchange. Not all journals have made such a decision. Some still publish only one kind of article, perhaps two. Science has much room for different visions of what a periodical should be and do.

But the point is this: every type of published writing in science is a distinct contribution to the field. Of course, it is common, and entirely understandable, that you might focus on getting your own research published above all else. Yet nonresults writing provides another avenue, at times more relaxed, for playing a significant role in your discipline. It is another opportunity for adding valuable content and (let us not overlook) placing your name before peers and the public, including job review committees.

There is also the question of influence. The skilled scientific author, who publishes review articles, book reviews, and so forth, as well as research papers, is likely to have a larger professional presence. He or she will reach a greater number of readers and achieve higher levels of visibility and authority than the average scientist. The growing openness of many journals to different forms of writing has the effect itself of opening up new space for skilled writers to take center stage, to expand their contribution, increase their reputation, and gain status for their work. Print and online journals today offer more chances for writers to engage in productive self-interest. Indeed, even for those who would rather focus on research per se, some types of nonresearch writing—especially review articles and correspondence—can be used to further expose work already done or to pave the way for its fuller publication. No surprise, then, that those in charge of job decisions, whether in academia, industry, or the government, look upon such publications in a very positive light, and why scientists eagerly, and justly, list them among their published works.

What follows below are some basics regarding this "other writing," specifically its most common representatives: the review article, book review, and debate/critique (of another's work). Be assured that all of these types of writing, too, vary considerably between fields and journals. As with research papers, there are no universal standards—except the one advising you to explore what the journals of your choice have to offer. In every case, it will help greatly to identify and use models from the literature.

The Review Article: Function and Role

An essential part of every scientist's work is keeping up with his or her field, and, to the degree possible, related fields as well. This is no simple task, especially these days. But it is exactly what the review article is intended to help you do—summarize and evaluate recently published work on a specific topic of importance. Such a topic is usually broader than that of a research paper, for example, "Evolution: The Role of Human Influence" (instead of a single example) or "Aromatic Metal Clusters: Bonding and Stability" (rather than analysis of a particular compound) or "How Do Thermophilic Proteins Deal with Heat?" (not a discussion of *Protentis caloris* occurrence in a Galápagos sea vent). In many cases, a scientist is drawn to write a review article because, in the course of normal research, he or she has surveyed the literature in some detail and has some specific perceptions and opinions about it. Many scientists, in fact, carry around inside them review articles, like novels, waiting to emerge. The major difference, of course, is that reviews are easier to write and are likely to be more interesting and valuable to read (plus, you don't need an agent).

A good review article can do several things. It can (1) provide an idea of the current state of knowledge and, possibly, its historical development; (2) discuss recent and new directions in research; (3) point up gaps or limitations in this work; and (4) make suggestions for future research. Not all reviews, to be sure, do all of these things—some focus only on the first or second tasks. But some type of perspective still must be added. As such, this type of article is really the closest thing to a scientific essay, a form of intellectual appraisal. Indeed, in some part, what literary or art criticism is to the humanities, the review article is to science.

Over the past few decades, reviews have become invaluable to many disciplines, particularly in the life sciences. Keeping abreast of new developments is no longer a simple or easy task. The reasons are several: (1) growth in the number of journals, thus in the volume of published papers; (2) an ever-increasing variety of new disciplines; (3) deepening specialization within individual fields; and (4) an expanding trend toward multidisciplinary research. These realities have everything to do with how science itself is advancing in the present age. The effect, however, is to make it more challenging and necessary to stay current on new thinking and discovery. I say this not to suggest the reader feel bound to a hopeless situation (on the contrary!), but instead to highlight the practical importance of the review article. While reviews are more common in the biological and medical disciplines than in the physical sciences, this seems likely to change, for all the reasons given above.

As with research papers, the review article comes in different shapes and sizes, depending on the publisher. Some journals publish reviews up to 20 pages long, meant to survey a topic and discuss where it might, or should, be headed. Other periodicals offer "minireviews" of five pages or less, intended to hit the moving target of a fast-paced area. Still other journals provide both types. There are journals devoted entirely to reviews. But reviews are also collected in book form, as in the well-known *Annual Reviews of . . .* (medicine, ecology, geology and geophysics, etc.) and various yearbook series. Thus, as with any other type of writing you may be interested in doing, by all means explore your field.

Reviews Are Commissioned

Review articles are usually commissioned. This doesn't mean that only the editors and their immediate family are ever able to contribute. It means that prospective authors need to submit an article proposal for consideration. An effective proposal usually contains a topic synopsis, an outline (annotated in most cases), and a description of the number and content of any figures. If the editor accepts your proposal, he or she is requesting the article—in professional terms, it is "commissioned" or "assigned" The process is intended not to erect barriers, but to protect both editors, who must deal with considerable literary mass in any case, *and* authors like yourself, who might otherwise spend months laboring away on an article with little chance of acceptance. Having your proposal accepted, however, doesn't guarantee publication. There is the small matter of submitting a well-written, well-organized manuscript, which then goes through the same basic process of peer review and required revision as a research paper.

Your first step, even before launching into your proposal, should be to contact the editor to see if he or she is interested in your topic. This is the time to ask for any specific guidelines in submitting your proposal (form, specific content). Be assured: editors *always* appreciate being consulted in this way at the beginning. It is a form of professional courtesy, and as such, good diplomacy. Editors also understand that a review article can do two things for your career. It can establish (or confirm) you as an expert voice among your peers—no small achievement, in any field. And it can make you the source of controversy—because, in most cases, you are interpreting and even judging, whether directly or indirectly, the state of knowledge in your discipline. Such considerations mean that reviews are often carefully edited. As always, the ultimate goal is to produce a better article and thereby serve the readership.

Important Points for Writing a Review Article

As an essay of sorts, the review can be more fluid in its structure and style than a research paper. But for this very reason, it needs logic and flow at least as much, if not more. Most reviews have only two standard parts in the main body, an introduction and a conclusions section. Between these two endpoints, you have a great deal of latitude about how to organize your article, how many sections to use, what you call them, what you put in them, how long they might be, and so forth.

There are several basic approaches that authors use to write a review. Here are the most common.

Alternative 1. In the *historical approach*, you choose to outline the development of knowledge on a chosen topic. This may involve charting the emergence of a particular subdiscipline. To do this, you should discuss the milestone studies in chronological order and point up the more important findings, advances in method, and principles that have emerged from this work. Normally, this involves going back no more than several decades. However, if you are concerned with theoretical advances, with innovation or maturation of ideas on a fairly broad topic, you may want to review concepts that go further back than this. For example, on a new concept in evolutionary biology, you might find it helpful to even begin with reference to Darwin himself: "According to Darwin's original theory . . ." or "One of the key implications of natural selection is . . ."

Alternative 2. When employing an *experimental focus*, you concentrate on methods and materials, on recent trends in experimental work. This might involve answering such questions as, What aspects of the topic have been most avidly pursued? What methods have been used or developed to study the phenomenon, and what are their advantages and limitations? Have any problems been revealed in these methods? What forms of data analysis are used, and do they also have limitations (are there any important links between experimental methods and analyses of results)? What might be done to possibly improve research in this area?

Alternative 3. The *concepts-and-hypotheses* approach focuses on the state of knowledge, as intellectual content. What are the reigning hypotheses about the topic and who has framed them? How might these concepts relate to the larger theoretical framework in the field? To what degree is current knowledge explanatory, predictive, or descriptive? What is the situation of debate, as it currently stands? What positions have been taken, on the basis of what data? Have there been important, recent challenges to otherwise accepted concepts? Are certain reconsiderations underway? What significant questions remain unanswered, or not yet pursued?

Alternative 4. The *implications* approach seeks to discuss the latest developments and to outline their possible consequences. Questions here include, What are the primary advances with potential for practical application? What types of application might come from this new work? How might these advances improve existing technologies or apply them in new ways? What potential do they hold for new therapies, new and better predictive capabilities, or other applications? What important gaps or limitations need to be addressed in order that this potential be realized? How might this be done?

Alternative 5. Emphasis in *the future* approach is placed on the direction in which current research is heading and what this might mean for the future. Issues taken up might be, Where can we extend our present knowledge? What are the new and emerging areas of study? Are there major hurdles or limitations (other than funding) that must be overcome in order that progress be made? What problems have not yet been rigorously studied? Do we need to improve our research methods, consider new forms of data, expand our analyses, reinterpret previous results? Which recent studies have been more significant in pointing the way forward?

Obviously, these approaches overlap in a number of areas. Most reviews, in fact, combine aspects from two or more of them, but tend to adopt one as their principal orientation toward the subject. It usually helps, therefore, to choose an approach first and then begin to frame your outline around it.

A good way to build your outline is by writing down themes or perceptions you wish to offer—major points or ideas that seem relevant—and then searching for some logical order among them. As always, experiment, and then experiment some more. Put down more subjects than you might need, rather than less. You may have to tinker with your material (add, delete, rephrase) before an order begins to emerge. You may also find that, once these ideas have been put down and given a degree of organization, a different approach than the one you originally decided on now seems more appropriate. Don't become frustrated or defeatist; such changes are entirely normal and common. The better writer is flexible, adaptable to the material.

How to Begin

The introduction to a review sets up a basic plan for the rest of the article. It identifies the subject and its importance and outlines how it will be treated, what type of perspective is being applied. For this reason, it is worth dwelling on how to begin.

Begin large, at least fairly so. Use general terms to define your topic. This is one of the few places in technical writing where you can offer a broad, open-armed invitation to your readers. Here are a few examples I've come across:

> Explosive taxonomic radiations are very useful to the study of species formation. The extraordinary biological diversity of these systems is seen to evolve through multiple cladogenic events. (for a review on speciation in rapidly diverging systems, with lessons from a specific African lake)

> To survive, living organisms must be able to adapt to their natural environment. Nowhere on earth is this simple evolutionary principle more tested than in high-temperature waters. (on the subject of how thermophilic proteins deal with extreme temperatures)

> Our ability to perceive the dynamics of nature is ultimately limited by the resolution of our instruments. The history of optical instrumentation shows a remarkable advance in recent decades and places us at a new frontier, where ultrafast pulse generation by lasers is key. (on methods and results in the generation of ultrashort optical pulses)

It is perfectly acceptable, in other words, to open with a bit of philosophy, even drama. Different editors, of course, will allow this to different degrees: some may want you to be less grand, more focused on your specific topic:

> The large genomes of mammalian cells are vulnerable to an array of DNA-damaging agents. This situation requires that constant excision and replacement of damaged nucleiotide residues take place by DNA repair pathways, in order that potential mutagenic and cytotoxic accidents can be minimized.

Either way, it's good to create interest. Keep your first sentence short, therefore, and follow it up with a longer one that defines, or begins to define, your real subject. You might consider using questions to begin—questions, of course, that you fully intend to answer later on:

> What do we know about the origin of coalbed gasses? Where has this knowledge come from, and what are its implications for future energy resources?

Your main effort in the introductory section, however, should be to orient your reader as to the significance of your topic and the direction you're going to take in the review:

> Because of their exotic electromagnetic properties, hole-doped manganese oxides have stimulated considerable scientific and technological interest in recent years. . . . The unusual properties of these manganites challenge our current understanding of transition metal oxides and define a fundamental research problem involving study of the interplay between such factors as charge, spin, and orbital degrees of freedom. . . . Until recently, there was strong agreement regarding how ferromagnetic states are stabilized in these substances. However, new experimental results suggest that more complex ideas are now needed to explain some of the main properties of these oxides.[1]

Notice how these sentences develop the subject. First, they set up reasons for interest (aided by such terms as "exotic," "unusual"). Next, the nature of current research is described (this is followed by an explanatory paragraph in the original). Last, the main thrust of the article is revealed; we see what type of perspective the authors intend to add to the subject.

Once you've assembled such an introduction, the rest of the article should begin to fall into place. But—and this is important—be open to changing things as you go along. As I've said a number of times in this book, the process of writing is one of experiment and discovery. You may well find that, in working through your different sections, new ideas emerge, even about the type of perspective you want to give your material, for example, what are the most important limitations to current work, what directions might future research might take, and so forth. Since one of your

---

1.   A. Moreo, S. Yunoki, and E. Dagotto, "Phase Separation Scenario for Manganese Oxides and Related Materials," *Science* 283, no. 5410 (1999): 2034. Sentences here have been excerpted and slightly rewritten.

main efforts in writing the review is to survey a given topic and to say what it means, it's essential to be receptive to what the material can tell you.

Book Reviews

The triumph of the journal in science has not eliminated the importance of books. These still occupy a crucial position. Indeed, there are many kinds to consider: monographs, treatises, edited collections of research papers, transactions volumes, conference proceedings, reference works, not to mention books for more general audiences (many kinds here, from science writing to biography and history of science). Depending on the journal, all of these types may be seen as worthy of review, or just a few. Either way, book reviews now form a regular contribution to most premier journals. And the scientists who write them, and who write them well, provide a crucial service by adopting a certain mantle—they are judges of worth.

Book reviews are commissioned, like review articles. Editors often choose books for review and then assign them to individuals they believe are qualified in the particular field. But there is almost always room for outside suggestion. If you have a work you'd like to review for a specific journal, contact the editor and offer your services: give the title of the book, say why you think it's important to review, and state your qualifications for reviewing it. If your first choice of journal isn't interested, try another. Keep in mind that periodicals are constantly in search of good reviews—these types of articles are widely read and much appreciated by a journal's audience.

So saying, let me make a plea, a necessary one. When you write a book review, please write about the *book*—what's in it, how it's organized, how well it's written, how accurate and complete it's information is, how helpful or useful it might be and to whom. Does this seem obvious? But many, many reviews succumb to the temptation, here acknowledged, to wander off into other areas: issues (that the book raises), opinions (of the reviewer), debates, anecdotes, personal experiences, quotations, pet peeves, and so forth. Some of these ingredients are fine, indeed worthy, but *only* when they are woven into a discussion of the work at hand—otherwise, they are little more than pontification. The book is your true subject, first and last.

Book reviews may seem open, casual, and largely nontechnical in nature. Yet an effective review is actually a very tight-knit, even rigorous structure. It provides much information and critical evaluation in the space of a page or less, and may even do so in an entertaining way. Begin by briefly describing the subject of the book:

> The Karoo is an arid and semi-arid region that takes up as much as a third of South Africa and extends north and west into Namibia. It includes large treeless areas, desert, arid and moist savannahs, grassland and, in the most favorable spots, patches of forest. (for a review on a book dealing with ecological processes and patterns in the Karoo)

> Ideas regarding the origins of life have changed a great deal in the past two decades, as an ever greater array of disciplines have begun to tackle the problem. The subject no longer belongs to biology, but has been taken up by researchers in fields as seemingly disparate as astronomy, paleontology, and molecular genetics, with varied but often exciting results. (on a book about the molecular origins of life)

> In the dead of summer of 1620, Cornelius Drebbel offered to astound the English royal court by generating a room full of wintry chill. He succeeded, as we know from the writings of Francis Bacon, but how? We learn the probable answer to this in the opening paragraphs of Tom Schachtman's *Absolute Zero and the Conquest of Cold*.

You can use humor:

> A new textbook on evolutionary biology is likely to find the existing landscape crowded and full of impending extinctions. In order to survive, a text must either fill a specific niche or else range widely enough to cover all necessary ground. Remarkably, J. R. Harrison's *Evolution on Earth* is able to do both these things, and thus should provide teachers with the type of work that has long been needed in the field.

Each of these openings sets up a discussion of the volume at hand. What comes next is a description of the book's contents, followed by (or perhaps merged with) judgments about its accuracy, organization, utility, and so forth. This is a simple formula for writing a review—subject, contents, judgment—but it works, and is easy enough to follow.

Using models here is just as important as it is elsewhere in scientific writing. Keep a file of reviews you like. Read through them before writing your own. In studying their style and content, you may note some of the following points:

- A book review is different, and more interesting, than a book report—which is dry and boring ("This text is the second in a series on soil science ... Chapter 1 discusses ... Chapter 2 is about ... Chapters 3 and 4 take up ... I didn't like this book, because ...").
- When being strongly critical of a book, it's best to offer an example or two from the text (evidence) to support your judgment.
- Quotations, well-chosen, provide an excellent taste of a work under review (another form of evidence).
- The pronoun "I" is not usually appropriate to book reviews in science—you are supposed to be offering professional assessments, speaking with a voice of authority for the field, not expounding personal opinions. For these same reasons, the so-called editorial "we" is suitable to use only when you are quite sure you are speaking for most of your colleagues or for the journal.
- Some discussion should be given to any visual materials in a work, that is, figures, photographs, maps, and so on. What is their quality? How helpful and relevant are they? How well chosen?
- A book with too many flaws, about which you have only (or almost only) negative things to say, is probably not worth reviewing. Best allow it to lapse into obscurity on its own demerits. An exception to this is when such a work represents an important position in an area of ongoing scientific debate—here you need to discuss the issues perhaps as much, or even more, than the book itself—or perhaps when the book is written by an especially famous person. Make sure you check with the editor before starting on such a review.
- Remember that no book is ever perfect. The best reviews strive for balance. Impossible standards for judgment serve no one well. Every book that's worth reviewing has positive aspects—don't overlook mention of these.

Finally, be cognizant of your potential influence as a reviewer. This influence is real and not to be taken lightly. In providing (let us hope) a professional evaluation, you are effectively giving professional advice. Negative or positive reviews can strongly affect book sales and levels of readership. A highly negative review, on the one hand, is tantamount to don't-buy/don't-read advice and runs the risk of offending both the author(s) and the publisher, at least temporarily. On the other hand, a poor book on an important topic is tantamount to a wasted opportunity for everyone in the field, and this deserves to be pointed out.

None of this, in other words, is a reason to soft-pedal any strong criticism you feel is justified—and that you can justify in your review. It *does* mean that, if pressed, you can back up whatever you say with reason or evidence (or both), that any facts you call on are truly accurate, and—just as important—that your writing is thoughtful, not glib (criticism for the sake of mere cleverness or for the appearance of superiority does a disservice to everyone). In the end, book reviews are really complex documents of diplomacy. Weigh your words, so they might deserve the authority they will likely carry.

Debate/Critique

Debate is absolutely essential to science. It defines a process by which research results are often confirmed or denied and new problem areas are outlined. Debate is also how the profession acts to control the quality of its work. Many journals therefore carry a section devoted to critical discussion of previous articles. Typically, this will include one or more critiques of a recent paper and a response from the author.

Most scientists, as routine readers of the literature, are aware of both how necessary and how emotional such exchanges can be. Some instances are congenial, even friendly. Some are almost matter-of-fact or may have a coating of frost. But others are acrimonious. Fur can fly over the use of such terms as "erroneous," "misleading," "invalid," or "specious" when applied to one's lifework. Indeed, there exists an entire vocabulary (largely adjectival) devoted to scientific criticism, and researchers are well-advised to learn how to use it judiciously (and withstand it). Full discussion of this vocabulary merits a book of its own.

In every case, if you are the critic, consider both the purpose and the effect of the language you employ. Consider what it would be like to be on the receiving end. Your aim in offering critique is to help improve the science at hand, not to cast doubt on the personal credentials or abilities of individuals (except, of course, for cases of fraud, which need to be taken up in a wholly different way). Note the following alternatives:

In their recent paper on early hominid migration, Leibnitz et al. arrive at conclusions which are spurious and unsupported by any available data. Their interpretation of such migration as a slow, intermittent process runs counter to the results of nearly all current research in the field.

Leibnitz et al. state their conclusion that early hominid migration was a slow, intermittent process, not sudden and rapid, as other researchers have maintained. We believe this interpretation is flawed for the following reasons.

Here's another example:

The data provided by Lowell to suggest that clathrates exist near the surface of Mars are erroneous and misleading.

Lowell offers the intriguing hypothesis that clathrates exist near the surface of Mars. We question this conclusion and the data on which it is based.

And, finally, one more:

In their study on the eradication of *S. gallinarum*, Klinsmann et al. employed the methods of Barnes (1985), now widely acknowledged to be outdated and error-prone (see Johannson and Walters 1996). Such methods must be seen to invalidate or cast serious doubt upon the results of their study.

Klinsmann et al. employed the methods of Barnes (1985) in their recent study on the eradication of *S. gallinarum*, an important topic in epidemiological science. I would like to comment on these methods and their possible effects, specifically in view of the critique offered by Johannson and Walters (1996).

In each case, both alternatives achieve the same basic goal. This goal is to state the terms of challenge and set the borders and conditions of debate. The difference, of course, is that the second example avoids the implication that Leibnitz et al. are Neanderthals and shouldn't be allowed to practice science beyond the kindergarten level (thinking this is one thing; using words to convey its effect in print is very much another).

Professionalism in this context means keeping a cool head and a restrained voice. This is how you will best serve the field, the research at hand, and (not least) yourself. Like it or not, the words you use will inevitably project an image of yourself as a scientific professional: by keeping cool, you will appear more authoritative and worth listening to. Therefore, consider every use of a critical term carefully, for its effect, and back it up by hard evidence. Keep in mind that too many negative comments will cast doubt on your motives. The appearance of a personal attack will invalidate most anything you have to say. Thus, avoid writing anything in final form when in the throes of anger or some equally intense emotion (disgust, envy, contempt, love, etc.). Always remember that whatever you write, if published, will be there forever, for all to see. Frame your comments so that they focus on the subject matter, not on the "poor choices" made by its authors.

How to do this? Simple: first, jot down your major objections or concerns in the order they come to you, as you read and reread the particular article. Second, look these points over and consider which of them are the most important—you will need to be succinct, as editors demand this. Third, put these points in some type of order, give them each an introductory sentence ("Leibniz et al. further suggest that . . . However, this contradicts recent work by . . .") and fill them out with evidence. Your point-by-point discussion can follow the organization of the original paper itself or it can develop its own logic of challenge. Either way, be brief and to the point.

Similarly, if you are on the other side, respond to any such criticism in a like manner, that is, professionally in tone:

Anderson and Weiss raise some important points in their comment on our paper regarding early hominid migration patterns. However, their criticisms overlook several key factors.

In questioning the data and conclusions of my recent paper on Martian clathrates, Ardau et al. make the following problematic assumptions.

While it is true that the methods of Barnes (1985) have been questioned in recent years, many researchers have continued to employ them and, in fact, may be seen to have confirmed their validity.

Answer each point in turn, in the same order it was presented. Be concise and keep your tone even and controlled. Counterattacks ("Anderson and Weiss appear unable to understand the significant points of our paper") will damage your credibility and, in any case, may be rejected or amended by the editor (his or her credibility is also involved). Eloquence in this context of conflict resides in the appearance of intellectual civility and dedication.

Such, in fact, has always been the case. Let us reach back a bit, to an earlier episode of debate, surely one of the most acrimonious and protracted in the history of science. Indeed, it is a debate not yet settled.

"Mr. Huxley, was it through your grandfather or your grandmother that you claim descent from a monkey?" So, according to Huxley's own report in his letters, spoke Bishop William Wilberforce before a large public gathering, composed of scientists and laypeople, during an 1860 symposium on evolution held by the British Association for the Advancement of Science. Huxley's reply, you may imagine, is worth quoting:

> If there were an ancestor whom I should feel shame in recalling, it would be not an ape but a man—a man of restless and versatile intellect who, not content with success in his own sphere of activity, plunges into scientific questions with which he has no real acquaintance, only to obscure them by aimless rhetoric and distract the attention of his hearers from the real point at issue by eloquent digressions and skilled appeals to prejudice.[2]

Whether these were precisely the words Huxley spoke upon the occasion defines a target of some skepticism among historians and a topic that itself has been often debated. No matter. For our purposes, they remain a worthy exemplar. This is because Huxley showed that even in the midst of so public a forum, with everything to gain or lose, it is excellent form to compliment antagonists, even while destroying them with tact and aplomb.

2.   Quoted in L. Huxley, ed., *Life and Letters of Thomas Huxley* (London: Macmillan and Co., 1903), 1:272.

＊

11. THE PROPOSAL

It's hard to make predictions, especially about the future.—YOGI BERRA

Importance of Proposals

Contemporary researchers tend to be worldly individuals in at least one respect: they know that a proposal does not always lead to marriage. Indeed, in the often financially polygamous world of modern science, it is necessary for a researcher to seek many unions, some temporary, some long-term, all based on proper and well-expressed overtures.

The proposal is one of the more important technical documents written by scientists and engineers today. There are several reasons for this. The first, of course, is financial: proposals form the basis for evaluations that lead to funding (or not), and thus make a great deal of science possible. Second, writing a proposal forces you or your research team to step back and create a systematic plan for your work—to conceive and organize its activities, apportion responsibility, consider monetary needs and constraints, think about timing. Third, proposals very often serve as literary reservoirs: if done well, they can provide much material for future articles, for communication with peers, for talks and poster sessions, press releases, and so forth. Often enough, a proposal is the first real publication to emerge from a particular line of research.

Making a Case

What, then, *is* a proposal? How should we think of it, as a form of writing? Some authors have called it—and many researchers would probably agree, in feeling—a sales document. But this is not quite right. Writing that sells is not promissory in a true, contractual sense; it aims at getting around careful evaluation through rhetorical superlatives that focus on one-time exchange—bait to the unwary. If we think of proposals as somehow linked to the idea of marriage, the sales document becomes a literary vestige of "the second oldest profession."

In practical terms, then, a proposal is usually something more interesting, and complex. It is several things: a *request* (for interest and funding), an *argument* (for the significance of certain ideas), a *blueprint* (for work to be performed), and a *promise* (that the work will be done within specified limits). Let us take these in order.

As a request, the proposal is commonly written to be evaluated by two or more researchers in the same field. This means that other scientists, not an anonymous funding agency, are your audience. Your document needs to have sufficient technical information and to be written in an appropriate style. These scientists' time is valuable (especially to them); they are getting paid nothing or very little to do this gatekeeper work, which they take very seriously, so be as direct, economical, and to the point as you can. Anything you say wastefully can and probably will be used against you. In short, your proposal should have dignity—it must embody the professionalism of the evaluators and acknowledge the pressures they face.

As an argument, meanwhile, your proposal identifies and outlines a specific problem. In most cases, this means a gap in the existing knowledge that (urgently) needs to be filled. The importance of this problem, what else depends on its solution, is crucial to your discussion. You must convince your reviewers that your work will not only be worthwhile, but significant—that it will enhance the field (of which they are members) in some specific way. Nearly all the details presented in your document, then, should flow toward supporting this claim of significance.

As a blueprint, your proposal explains the work you are going to do to solve the identified problem. In simple terms, you have to show that you can create and present a reasonable, concrete research plan, that is, produce plausible timelines, budget analyses, equipment requirements, and so on. This often means, of course, that you are the maker of beautiful fictions (even fantasy), since it is always impossible (thank heaven) to constrain future research work in advance. Be assured: your reviewers know this. Most often, they will have generated proposals themselves and thus have dipped into the wax of "creative writing."

Finally, as a promise, your document is a pledge that you (or your team), personally, are serious, competent, and responsible, that you have thought things through and have the skill to carry out all work. This may sound self-evident, but the truth is that many proposals fail because they appear hurried or thrown together, don't follow directions provided by the funding agency (e.g., in the request for proposal), or are poorly written enough to give the impression that the researcher or research team just couldn't be bothered to take the time and care needed to produce a good document. This type of failure carries a promise of its own, of course (let us not speak its name), as well as a touch of insult. Yes, true enough, the hurried nature of many proposals reflects the conditions and means of their production—the dash to make deadlines, to invent details, to coordinate pieces by different groups, and to sew the whole together in Dr. Frankenstein fashion so that the result walks and talks without too many seams showing and without becoming a danger to its makers. Your proposal must indeed hide all this reality. But this is because the task is to guarantee that you will enter into a contractual relationship and abide by all the responsibilities involved.

In the end, therefore, a good proposal will make a case for both you and your work. As put by James Paradis and Muriel Zimmerman, it will reveal "the value of your idea, the elegance and good sense of your work plan, the strength of your preparation, the appropriateness of your facilities, and the economy of your budget."[1] And another point, too, is often overlooked. The proposal shows that you or your team can write. It reveals that, if and when the occasion arises, you will be able to generate good-quality articles or reports for the benefit of other scientists and therefore the progress of your field. A poorly written proposal on an exciting topic is like a beautiful face with one eye.

1.   J. G. Paradis and M. L. Zimmerman, *The MIT Guide to Science and Engineering Communication* (Cambridge, MA: MIT Press, 1997), 116.

Definitions and Realities

Proposals come in two basic types, solicited and unsolicited. You write a *solicited* proposal when you respond to an announcement that money is available from a particular source, such as a government agency, corporate sponsor, or foundation, and is to be spent in a specific area or on a particular topic. Such an announcement is known as the request for proposal (RFP) or, in the case of engineering contracts, invitation for bid (IFB). The RFP or IFB will nearly always provide guidelines, sometimes rather detailed, on how the proposal should be prepared. You write an *unsolicited* proposal, on the other hand, for potential sponsors that have not made any such formal announcement, but who have funding programs in place. In this case, the sponsor may or may not provide guidelines on how it wants a proposal to be written. Overall, solicited proposals enter you into direct competition with other scientists; unsolicited proposals do so only in an indirect way.

Researchers also need to understand the fundamental difference between grants and contracts. Basically, a *grant* is a form of financial assistance for approved work. A *contract* is a legal instrument by which a sponsor acquires (buys) certain services and resulting products. Grants, in theory, are like scholarships; contracts are much closer to purchase agreements. These differences may seem clear enough on the surface, but in fact they've always been a bit troubled, due in part to questions of ownership (e.g., when patents result from the research at issue) and how far such ownership extends. In recent decades, moreover, as the lines between "basic" and "applied" science have all but disintegrated in many fields, distinctions have become even more problematic. There are now certain crossover forms of sponsorship. One of these is the *cooperative agreement*, defined as a grant for work in which the sponsor (usually the government) will have significant programmatic involvement. Then, too, there are various *cost-sharing agreements*, in which a particular sponsor will put up only a specified portion of the total budget, with the remainder supplied either by the research institution alone or by a combination of other sponsors.

Obviously, given all this complexity, it is absolutely essential for you, the individual scientist, to become as familiar as you can with the types of research support common to your own field. Avenues open to the physical chemist are quite likely to differ from those available to the paleontologist or planetary astronomer. The institutional realities of professional science have become more heterogeneous—and more field-specific—with each passing decade, and there is every reason to think that this will continue in the future.

At the same time, however, there is one major trend that cuts across most fields. The past 40 years have seen solicited proposals come to dominate most scientific research globally and in the United States. This is largely due to the expanded role of government funding in science. Corporate sponsorship, meanwhile, has also grown, but is concentrated in certain fields, for example, biotechnology, computer science, energy-related research. In some countries, such as Japan, corporations account for nearly as much funding as the government, with both solicited and unsolicited proposals accepted. But in general, solicited science now rules over most of the industrialized world. This means that many nations have adopted or adapted the RFP system. Agencies, corporations, and foundations publish regular releases (e.g., annually or even monthly) announcing and outlining their research support. It is essential to gain access to these announcements, through hard copy and Internet subscription, if possible.

The rise in solicited science has brought with it certain effects. First, competition for available funds has increased. This means that reviewers are now almost routinely overwhelmed with the number of proposals they have to deal with, and are thus pressured to be skeptical, impatient, and decisive. The strong lesson here is that your proposal should be an easy read—clear, to the point, aimed at giving the reader maximum content in minimum space.

Second, the importance of government funding has brought an increased dependence on political trends (commonly euphemized as "national priorities")—evident, for example, in the United States, with such developments as the war on cancer in the 1970s, the so-called "Star Wars" research in the 1980s, and, in more recent times, the Human Genome Project, energy-related research, and the so-called BRAIN Initiative. An ability to be nimble in taking advantage of public, governmental, or commercial priorities (including hot topics, such as specific illnesses), and avoiding others that have collapsed, can be a real asset in acquiring research funding. The able proposal writer will keep aware of such trends in his or her field, for it often pays to attach one's own research—without too obvious a stretch—to them. Knowing which way the wind is blowing is a definite means of survival, as any meteorologist will tell you.

Third, government priorities also involve social considerations, such as diversity. In the last two decades, for example, many funding agencies have attempted to inaugurate grants specifically available to minority researchers and related institutions. There are also now grants for research that promises to have some positive effect on public understanding of science or on teaching and training.

The Internet has profoundly altered the funding landscape. All grant-giving agencies have websites where they provide essential information and, most often, application forms to potential grantees. There are key reference sites that collect information and links for many grant possibilities, such as the comprehensive site on US federal grants, Grants.gov. Most advanced nations now have such sites. Most universities and colleges now have offices of research support that provide online databases of grant opportunities. Nearly all grant-giving entities, government or otherwise, require online submission of proposals.

Merit Criteria

Scientists who review proposals are like editors. They must choose between competing entries, and they must do so under pressure. They are guardians of quality who are very much aware of this role and are deadly serious about it. Part of this seriousness involves the criteria they commonly employ to judge merit. Sponsors very often provide reviewers with such criteria, and, as responsible professionals, reviewers will certainly use them—but not without other considerations.

If you were to sit down (as I have, for the purpose of this book) and ask reviewers from a range of different fields just how they evaluate an individual proposal, what they look for to award high marks, you would find a list very similar to the following:

- *Scientific Content.* How good is the science involved? What questions or problems does the proposed research set out to solve and how knowledgeably are they presented? (Proposers must show adequate command of the relevant material.)
- *Significance/Importance.* Why is this research needed? How will it advance knowledge in its particular field? To what degree is it truly original or does it fill a specific "gap"? Where might it lead in the future? (The problem or question must be interesting, therefore defensible on grounds of significance.)
- *Feasibility.* Is the proposed research practical? Can it be performed such that the specific question or problem can be answered within the given

time period? (Many scientists can pose interesting questions, but some problems will inevitably be years away from being answerable.)

- *Clarity.* How well does the proposal read? How simple and straightforward is it? How often is it necessary to stop to figure out what is being said or meant? (Proposers must be able to express their ideas and information adequately to others.)
- *Flow.* How well is the proposal organized? How logically do the various parts fit together? Is there unneeded repetition? Are there gaps that force the reviewer to work harder than necessary to understand the proposed work or thought process? (The proposal should help the reviewer do his or her job easily, with a minimum of extra effort.)

The first two criteria will be found at the top of every sponsor's list of evaluation guidelines. Quality and importance of the science are primary: this is a given, across the board. The question of feasibility, meanwhile, is something that an experienced reviewer will always look upon as crucial, whether this criterion is specifically given by the sponsor or not (often it isn't). If the project isn't realistic, it probably shouldn't be funded: this, too, is a given. Thus, reviewers are concerned, first of all, with technical questions. But they are also influenced, to no small degree, by the type of reading experience a proposal creates, and this is where the last two criteria come to bear.

In recent years, some granting agencies have incorporated other merit criteria into their programs. The National Science Foundation in the United States, for example, and certain public support agencies in Britain and Europe, now ask that projects be justified in part on the basis of some of the following types of criteria:

- How well could the project help integrate research and teaching (teaching, training, and learning)?
- Will results be made available to a broad audience to help advance scientific and technological understanding among the public?
- Does the research involved enhance in any way the participation of minority groups?
- Are there any clear benefits of the proposed research to society at large?
- How interdisciplinary is the project, and does it help advance this type of research?
- In the case of grants given to scientists in developing nations, how well does the project strengthen or advance "endogenous" basic science, that

is, provide incentives to reduce the exodus of research talent to developed nations?

One or more of these criteria might apply to any particular proposal. As always, part of your responsibility as an applicant is to know what a specific organization is looking for. Some idea, beyond the details given above and those found in any agency proposal, can be gleaned from a visit to the organization's website.

In all cases, however, a good proposal is written not just to "get things down on paper," but to create real interest. Moreover, it should create such interest and leave it unsatisfied, in anticipation. If a scientific article tells a story, then a proposal does too, but without an ending. It will make the reader want to see the work done (and done well), to see how it all comes out. And this is where the quality of the writing itself comes in.

Reviewers see your proposal as an indication of how good a researcher you (or your team) are. If your presentation is cogent and logical, if it doesn't waste a reader's time with extraneous details or confused wording, you will appear in control of your subject. Furthermore, how well you respond to the specifications given in an RFP also tells a reviewer or sponsor something important: whether you can follow directions—and thus carry out the proposed work.

Thus, anything that might make your document clearer, more to the point, better organized, and precise is worth considering. As in all other forms of scientific writing, the process of revision is where the ore is refined into gleaming metal. Ask yourself, at every stage or step, whether a particular point or detail is really needed, what it adds to the total, whether it contributes toward the argument and toward the appearance that the work is important, well-planned, properly budgeted, and thus in the right hands.

Because it is a request for money, a proposal should project a sense of confidence, realism, and solid planning. Reviewers should be led to feel that any funding will go directly toward efficient completion of the project and related publications.

Use of Models (Again)

Let us return to one of the main themes of this book. With proposals, too, good models are invaluable, no less than in any other area of scientific writing (and perhaps even more, given what's at stake). One reason why models are especially important here is that scientific proposals come in an enormous variety of shapes and sizes, matching in a sense the varied institutional dependencies of science itself. In a number of fields, including my own (geology), they range from brief, 10-page documents to research bids that stretch to several volumes and thousands of pages. Becoming intimate with proposal formats in your own area qualifies as a necessary form of literacy.

Choose examples to study and imitate that were not only funded but that received high marks from reviewers. Veteran colleagues are one of your best resources here. Ask them to supply you with samples they consider particularly worthy—then ask them why they consider them so. Such samples might include proposals these colleagues wrote or worked on themselves or others they reviewed (many scientists keep copies of proposals they've worked on or given high marks). Use common sense in choosing your models. Be sure that they are recent—proposal requirements change over time—and that they targeted sponsors you yourself might select.

Also think about collecting one or two examples of proposals that were poorly done—where the problems were clearly identified and commented on by reviewers. Learning what to avoid, or at least watch out for, can be as useful as finding what to emulate. Or, in more blunt terms: nothing succeeds like avoiding the errors of others.

In studying your positive examples, pay attention both to broad points and to details. Broad points include, How much knowledge of the specific topic was assumed? How are the main goals laid out? How is the reader urged to see this as a truly significant and original project? More detailed points to consider might be, How long is each section? How is it organized? What type of style was used; for example, did the author employ questions, subheads, enumeration, or other aids to guide the reader? What specific parts of the proposal seemed to you especially strong, well-written, concise?

Besides studying good and bad examples, you can learn a great deal from experienced colleagues through conversation. As noted, colleagues who are experienced proposal writers, and who have served as reviewers, are a great resource. Sit down and talk with them about their perceptions. Ask them what impresses them about a proposal, what common blunders they see, what makes them thankful or frustrated, what points they would stress were they to give a lecture on the subject. Chatting about the whole business can bring up practical tips that might otherwise not be evident from the written example alone. Most important of all, use your colleagues as first-pass reviewers: have them read through your own proposals and make suggestions.

Finally, keep your eyes and mind open. Whatever models you choose to start with, consider adding new ones at a later time if you come across any that seem better or more appropriate. Writing, as I've said many times in this book, is an experimental process, which means a process of continual learning. As your proposal-writing skills advance, your needs for improving and refining these skills will naturally change. Good writers are always interested in adding new tools to their workbench.

Example

The succeeding pages of this chapter offer portions of a well-written proposal that gained high marks from reviewers and was therefore funded (by the National Institutes of Health, a major research agency in the United States). It concerns research in the field of immunology and is highly technical. The alert (or even not so alert) reader will note that I have retained a significant amount of the original vocabulary—this is done for specific reasons. First, most scientific proposals *are* highly technical, indeed they *have* to be, thus any other type of example here would be superfluous. Second, it is very important to be able to "read through" this terminology in order to understand the actual patterns of language flow and the logic involved. Only parts of several sections are included (the original was over 30 pages long).

To begin, let's look at the description section (fig. 11.1). A majority of proposals today ask for this type of abstract at the beginning and may specify what it should include: long-term objects, specific goals, basic research design and methodology. As shown, the abstract must fit into a given space, which demands that it be no more than about 400–500 words, and often less.

Note in the example how the very first sentence presents a short, concise statement of the overall goal. There is a sensitivity to the reader here—a simple, straightforward entry into the project. This is then followed up by an enumerated listing of specific aims, surveying the What of the proposal. Next comes the How, a brief discussion of approach/methodology. Finally, we are told the Why—first in terms of specific significance ("discovery of genes involved in class I and class II restricted antigen processing"), and then related to more general importance ("defenses against

DESCRIPTION. State the application's broad, long-term objectives and specific aims, making reference to the health relatedness of the project. Describe concisely the research design and methods for achieving these goals. Avoid summaries of past accomplishments and the use of the first person. This description is meant to serve as a succinct and accurate description of the proposed work when separated from the application. If the application is funded, this description, as is, will become public information. Therefore, do not include proprietary/confidential information. **DO NOT EXCEED THE SPACE PROVIDED.**

The overall goal of this grant is to identify new genes involved in antigen processing and presentation. The specific aims are: 1) to search for genes that affect the processing and presentation of HLA class II restricted antigens; 2) to search for these genes in the MHC, in unstudied flanking regions of chromosome 6p, and in the rest of the genome; 3) to identify new genes that affect the processing and presentation of class I restricted antigens; 4) to investigate whether the MCH-linked heat shock genes have a role in antigen processing and presentation; and 5) to investigate the basis for the alteration in recognition of antigen processing/presentation mutants as targets for alloreactive T cell clones. The basic approach to new gene identification will be to isolate mutants affected for processing/presentation, using selective schemes designed to distinguish between antigen processing-competent and -incompetent cells. . . It is anticipated that these studies will lead to the discovery of genes involved in class I and class II restricted antigens processing. In a more general way, this project should enhance our understanding of the mechanisms involved in antigen presentation to T cells, mechanisms that are critical in host defenses against microbial pathogens and that may play a role in the pathogenesis of autoimmune diseases.

PERFORMANCE SITE(S) (organization, city, state)

KEY PERSONNEL. See instructions on Page 11. *Use continuation pages as needed* to provide the required information in the format shown below.

Name   Organization   Role on Project

Number pages consecutively at the bottom throughout the application. Do *not* use suffixes such as 3a, 3b.

11.1   The abstract portion of a grant proposal

microbial pathogens" "pathogenesis of autoimmune diseases"). The whole thus forms a neat closure, moving from What to How to Why, and at the same time starting with the general, moving to the increasingly specific, and then advancing outward to the general again. This type of hourglass approach (broad, narrow, broad) is among the most effective rhetorical techniques for guiding readers through any project, and for giving them the sense that they have grasped the essence of the matter.

The budgetary details show a common type of breakdown for direct costs (fig. 11.2). Note that the requested salary amounts are calculated, essentially, on the basis of person-months: this is typical for many proposals in the United States and elsewhere. Almost all proposals will ask you to break out specific equipment and supply costs, as well as any outside consultant or service (e.g., computer processing) costs. It is also expected—and indeed, might be advised—that researchers include costs of travel to conferences where they will present the results of the funded research, and also page charges for publishing this work.

Now consider the research plan. Here is where the authors flesh out details regarding the aims, background, and significance of their project. In the example, the specific aims are listed, as they are in the description (fig. 11.3). The wording is not exactly the same in both—note that a bit more detailed information, including terminology, is given here, especially in the last two items. But the same order is kept, indicating that the writers are aware of their earlier pattern and presentation. Moreover, the degree of specificity tends to increase toward the bottom of the list—again, the pattern moves from more general to particular, which is an excellent way to guide your reader into the matter of your discussion.

This same pattern emerges again in the next section of the plan, on background and significance. Here is one of the most important parts of any proposal, since it presents an argument for the rationale, intelligence, and ultimate value of the project. Succeed here and your readers are your allies. Notice that, at the very beginning, the authors outline the gap in knowledge they will try to fill. They do this, moreover, by stating what is known and what is not known, first very generally (in the initial few sentences), and then, in more detailed fashion (second paragraph). The use of questions is nicely done: not only do they draw the reader in, emotionally and intellectually, but in their particular order, these questions offer a chain of logic that suggests a well-reasoned approach to the research at hand.

Finally, in the third paragraph, the significance of these questions is brought back out to the larger realm. The authors place their thinking in

Principal Investigator/Program Director (*Last, first, middle*):	Newton, Isaac		

| DETAILED BUDGET FOR INITIAL BUDGET PERIOD DIRECT COSTS ONLY | | | | | **FROM** 9/01/2002 | **THROUGH** 08/31/2003 | |

PERSONNEL (*Applicant organization only*)					DOLLAR AMOUNT REQUESTED		
NAME	ROLE ON PROJECT	TYPE OF APPT. (months)	% EFFORT ON PROJ.	INST. BASE SALARY	SALARY REQUESTED	FRINGE BENEFITS	TOTALS
Isaac Newton, LLD	Principal Investigator	12	24	125,500	31,375	7,450	38,825
Robert M. Hooke	Co-Invest.	12	50	96,500	48,250	8,200	56,450
John R. Keill	Sen. Fellow	12	75	65,500	49,125	5,100	54,225
Samuel Clarke	Sen. Fellow	12	50	60,000	30,000	3,750	33,750
Gottfried W. Leibnitz	Research Tech.	12	100	30,500	30,500	1,250	31,750
Subtotals $\longrightarrow$					189,250	25,750	215,000

CONSULTANT COSTS	
EQUIPMENT (*Itemize*) Centrifuge ...... 11,000 Spectrometer 8,000	19,000
SUPPLIES (*Itemize by category*) Serum .... 9,500 Synthetic oligonucleotides .... 1,500 Tissue culture media .... 2,100 Isotopes, labeled compounds .... 5,600 Chemicals, enzymes .... 5,800 Small equipment items and X-ray film, filters .... 1,300 misc. supplies .... 3,794 Glassware, plastic .... 4,100 Liquid nitrogen, gases 1,800	35,494
TRAVEL Two trips to national meetings by PI and one co-investigator	6,500
OTHER EXPENSES (*Itemize by category*) Page charges .... 2,500 Books, journals, software .... 2,000 Protein sequencing .... 1,000 Equipment maintenance .... 2,500 Use of Flow Cytometer .... 2,000 Telephone, postal 500	10,500
TOTAL DIRECT COSTS FOR INITIAL BUDGET PERIOD $\longrightarrow$	$ 286,494

11.2   The budget section from a grant proposal

the context of ongoing work—they admit their debt to others and, at the same time, include themselves in the community of researchers on related topics—and state that what they are really after is new knowledge that will advance "our understanding of immune system response," especially in the thymus gland.

RESEARCH PLAN

A. Specific Aims

1. To search for genes that affect the processing and presentation of class II restricted antigens, specifically in portions of the major histocompatibility complex (MHC) and in unstudied flanking regions of chromosome 6p.

2. To search for genes, affecting class II restricted antigen processing and presentation, which map elsewhere in the human genome than on the short arm of the 6th chromosome.

3. To search for new genes that affect processing and presentation of class I restricted antigens.

4. To evaluate what role, if any, the MHC-linked heat shock genes and their cognates may have in antigen processing and presentation.

5. To investigate the basis for the alteration in allorecognition of antigen processing/presentation mutants by alloreactive T cell clones.

B. Background and Significance

Although researchers have gained many important insights on the biology of antigen processing and presentation, little is yet known about the specific genes and gene products that carry out these functions. This is particularly the case for class II restricted antigens, which have a crucial role in immune system response. Prior research has confirmed, however, that the overall process is quite complex [references]. At a minimum, it involves several main aspects, including: the biosynthesis of class II α and β and invariant chains in the endoplasmic reticulum; the association of these chains in a nine-unit complex; and the trafficking of this complex through the Golgi to the trans Golgi reticulum, directed by an Invariant chain signal, and thence to the endocytic pathway, where the Invariant chain is degraded. . .

While this scheme is probably correct in general terms, many of the details remain unclear. For example: Where and how do peptides derived from lysosomal processing become associated with class II molecules? Do class II molecules recycle through the endosomal compartment and acquire newly generated peptides for presentation, and, if so, in what cells? In what compartment are endogenous membrane proteins degraded to yield the peptides which associate with class II molecules?

Work from several labs, including our own [references], suggests that the answers to these questions are very closely related and will have an important effect on our understanding of immune system response. In particular, the peptides that derive from endogenous proteins constitute a major class occupying the binding grooves of cell surface class II molecules, and appear involved in positive selection and the induction of self-tolerance in the thymus.

CONTINUATION PAGE: STAY WITHIN MARGINS INDICATED

11.3 The research plan for a grant proposal

All of this, together, offers an auspicious beginning. It is clear to a reviewer that the proposers can do several important things: they can think through their ideas clearly, creating a solid, rational plan for them, and they are skilled enough as authors to generate good-quality publications from the relevant work. Their claims are not too grandiose, nor are they overly modest. They project confidence, competence, and realism. Much of the rest of the proposal—including a section on research design and methods—takes up the task of demonstrating that the relevant work is feasible, and can be done within the budgetary limits presented. A reviewer is left with the feeling (however imaginary) that any funding will go directly toward efficient completion of the project and related publications.

Finale

In the end, there will always be a fictional aspect to proposal writing. How could it be otherwise? Science is involved in many marriages, with institutions, social trends, political realities. Proposals are a way of admitting these dependencies by telling plausible stories about the future. These are stories we tell to fellow travelers, other scientists. They are different from those related in articles, reports, and presentations; they have their own demands for believability—demands that are linked to the practical realities of science in the world today. To advance knowledge, in part, means convincing others that our work has been, and will be, valuable, well-conceived, and worthy. This, in itself, would seem a worthy skill to cultivate.

12. GRAPHICS AND THEIR PLACE

It is only shallow people who do not judge
by appearances. —OSCAR WILDE

Visual Language: Separate but Equal

Modern science relies deeply on illustration—graphs, charts, drawings, photographs, maps, models, and other forms. Technical knowledge today is inseparable from visual presentation, from its specific powers to order and convey information. Scientists, moreover, appreciate excellent graphics. Illustrations that offer data with clarity and elegance are a unique type of achievement—creative, efficient, even a source of delight.

Scientific illustration has an extremely rich and venerable history, reaching back to ancient Egypt, Greece, China, India, and medieval Islam. Alloys between science, art, and draftsmanship, forged in the European Renaissance and after, are still evident in many aspects of contemporary image making—in drawings of specimens, attention paid to form and balance, three-dimensional effects, uses of color. Many of the most influential works in the history of science—Galileo's *Sidereus nuncius* or Vesalius's *De humani corporis fabrica*, for example—have been books of pictures as well as text.

That being said, it should be stressed that the visual dimension to science forms a language all its own, a kind of pictorial rhetoric, if you will. By this I mean that graphics are often much more than a mere handmaiden to writing. They don't just restate the data or reduce the need for prose, but offer a kind of separate "text" for reading and interpretation. To assure yourself of this, take any well-illustrated article, copy the figures, and assemble them in order of appearance. You will find that they tell their own story, in some manner parallel to that of the writing, but in other ways different, enriching, though also with notable gaps.

Illustrations serve a variety of functions. Charts summarize data and make comparisons. Graphs provide analysis by revealing patterns, relationships, or possible correlations. Images, meanwhile, offer different kinds of evidence, explain and explore information, demonstrate specific points, represent concepts or theories. All in all, this is an impressive array of service—and it certainly helps point up (and justify) why scientists often browse through articles by reading the abstract and looking at the illustrations.

Perhaps most fundamental of all, however, visual discourse adds variety for the eye and enhanced appeal for the mind. Does this seem trivial? It shouldn't: the psychology of reading is not a little complex. The living brain very much appreciates intelligence expressed in different forms.

Specificity and Change

As a scientist-author, fresh on the trail of publication, it's a good idea to become intimately familiar with graphic elements used in your field. This may seem obvious. But there are two factors that raise it beyond mere common sense.

First, many visuals are highly field specific. Even graphs, maps, or other presumably standard figures can change considerably in form and style, as well as content, between disciplines. Moreover, there is the "journal effect" to consider: even within your own field, different periodicals have their own demands for how they want articles to look, just as they have standards for written copy.

A second reason to get familiar with graphics in your area is that many types of visuals are undergoing change, due mainly to the advent of digital technology, and will continue to evolve as this technology does. Indeed, the digital age has introduced a fertile array of new visual possibilities: satellite imagery, three-dimensional modeling, ultrasound technology, tomography, magnetic resonance imaging, various types of electron microscopy, and a dozen others. Science is replete with new powers of vision, new means of making the informed eye the instrument of study, analysis, and discovery. Color, too, has found an expanding role in scientific imagery and continues to revise older forms. Those who grew up in science before the 1990s can testify to a former universe of black and white, where the tones of text and image were identical. Since then, a great change has taken place, with color now abloom in every field, being much more easily and routinely and cheaply achieved in the newer online cosmos. More than ever, pictures today often demand and reward attentive study in themselves.

Becoming familiar with the graphics of your field is therefore essential—but it is not enough, in and of itself. This new sophistication depends on software, which is also often designed for individual fields. Certainly, a number of generic programs exist to help scientists create graphs, charts, and tables. Most major software vendors offer programs that cover basic graphing capabilities, useful in many circumstances when simple plots will do. For more complex demands, there are programs designed specifically for scientific uses (SigmaPlot and Prism are two widely used examples, but there are others too). Beyond these general-purpose programs, software is likely to be specialized. Fields as diverse as molecular genetics, petroleum geology, climatology, and mechanical engineering now each have a large spectrum of dedicated graphics programs to use with particular types of data. This places new demands on the scientist, who often must learn at least some of these programs.

Finally, new visual possibilities are attached to Internet communication. Still very young, Internet science promises new types of nonwritten expression: real-time animation, interactive modeling, use of video and audio. In some areas, these capabilities have already been taken up—online journals in medicine and chemistry, for example, have included animated graphics, video, and interactive visuals of various types. Broadband capabilities make for new possibilities in scientific illustration. Yet it's important to remember that broadband speeds are highly variable around the world; they are much less, and less available, in large portions of Latin America, Africa, Central and South Asia, and Southeast Asia. This may be something to consider if you (or your research team) are aiming for a fully international audience. Visuals of very large file size may have a more limited audience, at least for now. Broadband, too, will continue to evolve and expand in most places.

All of which highlights, again, the need for the individual scientist to learn what is out there, both in terms of what is being done and whom it can reach. Becoming an effective author means becoming literate in these visual forms of communication—learning how to read them, how to recognize good *and* bad examples, when and where to imitate the best, and who your audience is likely to be.

Choosing Models: A Help Here, Too

As with writing, you can learn a lot about producing good graphics by studying the admirable work done by others—and the opposite. Not only will this provide you with guiding examples to emulate and avoid, it will also help sharpen your critical faculty about what goes into such an image, what makes it effective, easily deciphered, informative, attractive. You can collect entire articles or only one or two illustrations from a paper. You might want to make a fairly large collection at first and then whittle it down to the very best: the exercise of doing this will decidedly focus your attention to details of quality. You might try gathering several nice examples of a single type of image, compare them side by side, and choose from there. Any method that helps you scrutinize and evaluate the images of your field is valuable. Find out what frustrates or irritates you (Too much type? Text too small? Figure too busy? Poor labeling?). The best questions to ask, as always, are, Is this something I wish I had done? or What would I do to make this better?

For every one of your choices, or for any graphic you find particularly worthy (or the opposite), stop and ask yourself what it is that seems especially good (or poor) to you. The more you come to understand your own preferences, the more you can use them consciously in preparing your own articles. Here are a few categories to help you judge individual illustrations and analyze your impressions:

- *Neatness.* Is the image clean and sharp? Does it invite attention, or repel it?
- *Readability.* Can the eye move over it and pick up information, either quickly or with concentrated attention, or is there too much confusion, too much data (a "crammed" feel), a lack of integration?
- *Use of Type.* Is the font easily readable, large enough? Is it placed well, or does it invade and distract? Is there too much of it (a common error)? Is there a proper hierarchy among type sizes—that is, do the largest words refer to the most important items, the smallest to the least important?
- *Size.* Is the image too small to be fully visible? Do you find yourself drawing the page nearer to your face? On the other hand, does it fill the rect-

angular space to the edges, as it should, or does it have too much white space?

- *Aesthetics.* Is the image balanced, or does it seem lopsided? What aspects or portions draw the eye most, and are these significant in terms of content? Are the thickest lines on the graph the most important? Do the patterns in a bar chart highlight the differences you want to show?
- *Use of Color.* Does the use of color help distinguish content, increase readability? Are colors appropriately distributed (e.g., blue for depth, red for height)? Is text minimized yet visible? Is the legend (if needed) simple and easy to use? Are colors consistently coded among different images?
- *Consistency.* Do similar images (maps, charts, graphs, photographs, etc.) carry the same stylistic scheme in terms of line width, type font and size, labeling, scales, and so forth?
- *Room for Improvement.* Are there any changes you yourself might make to improve the quality of this image?

Experiment, Experiment

I said earlier that writing involves experimentation—trial and error, revision, working things out. This is equally true for figures. It is true even for seemingly simple visuals, such as charts or graphs. We often construct these from our data as tools to help us in our analysis, to make comparisons, look for trends, discover relationships. When we initially draft our illustrations, we are frequently performing what are akin to visual trials—recasting our information in new forms to see if something important and unforeseen steps into the light.

The experienced scientist-author knows, moreover, that it can be very helpful to take a single data set and graph it in different ways. This means trying out distinct analytical versions in order to discover what form offers the most effective, meaningful presentation. Today this can often be done for charts and graphs, in particular, with a few clicks of a computer mouse: histograms can be changed to line graphs, pie charts, dot graphs, and other forms, with or without labels, error bars, means, and averages. Current software gives you the power to play with data in productive ways. As with text, the process of creating any specific image can reveal new aspects or relationships previously unnoticed.

So it is entirely normal—indeed it should be expected—that you'll often create more illustrations than you need, and that you'll need to revise the ones you keep. Sometimes a particular graphic, one you may have worked hard on, will be unnecessary; its message can be stated in a sentence or two of text—if so, please (for the sake of scientists everywhere) delete it, grieve briefly, and move on.

Experimenting with graphics also means making decisions about appearance. In a good scientific illustration, everything visual qualifies as content. Choices therefore need to be made about such things as how much text should be used; what type font and size are best; what shadings or colors are appropriate; what kind of scale should be used; how far each axis should extend; how wide the bars or lines should be on a chart or graph; what patterns should be used; whether a legend is needed, and where should it go; and so on. For more complex figures, the decisions are likely to be even more numerous.

Luckily, the relevant answers to such questions are far from open-ended. There is much standardization in scientific imagery, and your models will help guide your choices. No example, however, is ever a final template. There will also be factors (let us hope) individual and original to your work, and these will require that you adapt particular graphical forms to your specific case. This may involve changing scales, altering colors used in a model, graphing an added variable.

Think of each figure, therefore, as a draft in the beginning, a kind of visual audition. No time spent trying things out intelligently is ever wasted. It is sometimes necessary to discover what graphical forms might work best with your data, which types of illustrations you are most comfortable using. After all, this too is an area where authors develop a certain style, however subtly expressed. Don't berate yourself for any dead ends; be thankful when you find and overcome them. Experimentation, in both writing and illustration, is one of the most crucial processes in learning how to communicate well.

Many articles need (but often don't have) an introductory illustration to help orient the reader. Could your document benefit from such a graphic? The answer might be yes, no, or maybe, but it's usually worth asking and thinking about the question. An introductory graphic can serve many of the same purposes as the introduction itself—it can provide setting, such as can be done with a map, large-scale model, and so forth; it can offer essential background, for example, in the form of a flow chart that shows a sequence of relationships, a schematic diagram of apparatus, a time-based graph or chart outlining the relationship under study. Again, a guiding principle is to think of what might help usher your reader into the domain of your investigation. What would you use in an oral presentation to do this?

Try to make sure that any diagrams and drawings are relatively pleasant to look at. Use a type style that can be read very easily. Lowercase lettering is usually more pleasant to read than all uppercase. Many journals prefer a sans serif font for illustrations: Arial and Helvetica are common choices. Others prefer serif styles or are nonspecific. Check this out *before* you create your images—spare yourself (or your artist) the agonies of unnecessary revision.

As with text, different journals will have different specifications for many aspects of the illustrations they agree to publish. At a general level, this is likely to involve such aspects as software format (preferred programs), sizing, method of delivery (whether via the Internet or as printouts). Some journals, however, are quite specific about the detailed form your graphics should take, for example, type style and size, line thickness, use of borders and arrows, labeling (where and when), scale bars, and more. Most periodicals spell all of this out in their instructions to authors. Some demand consistency of stylistic detail among articles, and some are less exacting. Thus, you really do need to take a close look at published examples in your journal of choice before spending the time to design and create your images.

Some Necessary Pointers

First, *keep text on your figures to a minimum.* Use it mainly for labeling, not for explanation (leave this to the caption and main text). Also, be consistent from one figure to the next in the fonts you use, numbering and lettering style, and other such aspects.

Avoid any overly fanciful or arcane fonts for your images. This will only distract the viewer (while drawing attention to yourself as the possible embodiment of bad or eccentric taste).

When needed, use different font sizes to indicate different levels of importance. Be consistent about your sizes and font styles from one graphic to the next. Alternatively, consider using boldface or italics as highlights or as a means to add a dash of visual interest. Be aware, however: too much variation is distracting, so keep your visual hierarchy simple, clean, efficient.

It is common practice for many journals to offer readers the option of downloading a PowerPoint slide (for teaching) of most or all of the figures in a paper. If this is the case, you must design your figures appropriately. This means they need to be clear, sharp, and readable both at a reduced size *and* when blown up on a slide.

Make sure all illustrations, whether originals or scans, have a resolution of 300 dpi (dots per inch) or higher. This is the absolute minimum required for printing; 600 dpi is usually better. Provide high-quality laser printouts of every figure, if these are requested.

Use a simple, clear system for naming digital files of your artwork. Common schemes include surname of the lead author + figure number (Montgomery.2); one or two keywords + figure number (Apoptosis.2); and simple abbreviation of keywords + figure number (CrysRNA.2).

Finally, remember that *true elegance in science resides in simplicity and restraint.*

Examples

In the pages that follow, I've tried to present a series of good and bad examples of illustrations taken from the published literature. In each case, a brief commentary, and possibly a question or two, are given in order to help you evaluate the relevant image and thus further your own critical ventures.

Charts and Graphs

Charts and graphs are particularly well represented in figures 12.1–12.10. This is because these graphics are surely the most ubiquitous in all of science. My own unofficial survey of more than 150 journals in 57 fields (from insect physiology to mathematical physics) shows that they occur almost twice as often as any other visual genus. Indeed, in some periodicals, they are the *only* graphical form to be found. Thus, their importance comes in rather high.

Tables present exact numerical data, whereas charts and graphs take these data and give them a visually analytical form. Use tables when you need to show specific or precise values; use charts and graphs when you want to find and express meaningful relationships from these numbers. Very rarely will you ever need to show both.

Notice in the figures that follow that, in most cases, the horizontal axis

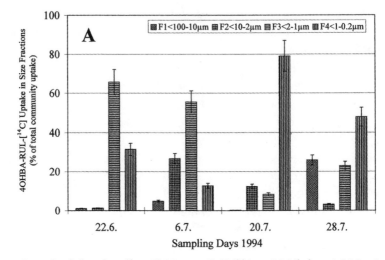

12.1   Example of a bar chart (from U. Munster, E. Heikkinen, M. Likolammi, M. Jarvinen, K. Salonen, and H. De Haan, "Utilisation of Polymeric and Monomeric Aromatic and Amino Acid Carbon in a Humic Boreal Forest Lake," in "Advances in Limnology," special issue, *Archives of Hydrobiology* 54 [1999]: 118; used with permission of Schweizerbart Publishers, www.schweizerbart.de)

plots an independent variable (the data set that we select), whereas the vertical axis gives the dependent variable (what we measure).

CHARTS

Bar charts or graphs, particularly histograms, are extremely common. The essence of such charts is usually to make comparisons between data sets. They are thus particularly useful for showing differences, and are less suited for revealing trends or relationships.

Take figure 12.1. This shows the uptake of a particular carbon-bearing chemical by a forest lake, during a series of sampling-day intervals. Several aspects of this chart seem well done. First, the axes are labeled in type large enough to easily read, and a legend is provided. Second, the bars are of adequate thickness (this is not trivial) and show appropriate error ranges. Third, the different data sets are properly separated. Fourth, each data set shows the same order of bar patterns.

What can be done better? The patterns for individual bars are much too similar to be easily distinguished. Note, too, that the legend is too small

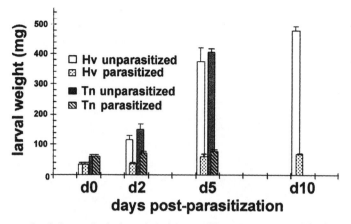

12.2   Example of a box-and-whiskers chart (reprinted from L. Cui, A. I. Soldevila, and B. A. Webb, "Relationships between Polydnavirus Gene Expression and Host Range of the Parasitoid Wasp," *Campoletis sonorensis: Journal of Insect Physiology* 46 [2000]: 1401; copyright 2000 Elsevier Science; used with permission)

and cramped. These problems leave the data difficult to decipher and thus interpret. The *x*-axis is also improperly labeled: shown are sampling *dates* (day, month), not days. Solutions to any and all of these problems are simple and straightforward.

Now examine figure 12.2. This box-and-whiskers chart is an attempt to show the effect of parasitism on weight gain in two species of butterfly (*Heliothis virescens* and *Trichoplusia ni*, abbreviated "Hv" and "Tn" in the figure). Nearly everything is clearly labeled on this figure; the type is large and welcoming. Bar patterns are easy to distinguish, the legend is neatly done, and the interpretation of the data is clear (parasitism halts growth at an early stage). Moreover, the author has nested the bars for each individual species to highlight the comparisons being made.

Could anything be improved? The whiskers sticking out of the top of each bar could be explained (errors bars?). The intervals on the horizontal axis could be a bit wider, too, allowing for a widening of the bars. d10 could be placed in its rightful interval, rather than in d9—which suggests that perhaps two-day intervals (d2, d4, d6, . . . d10) might have been a superior choice for sampling. Finally, there is a larger question: would the data be more revealingly shown in the form of a line graph? If the authors' point were to show only gross, overall patterns, then the answer would be no (or rather, not necessarily). But if ideas of developmental progress are at

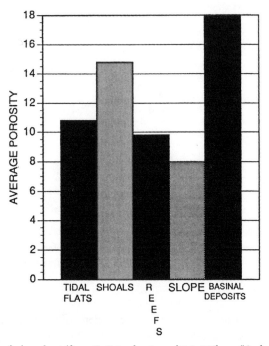

12.3   Example of a bar chart (from C. F. Jordan Jr. and J. L. Wilson, "Carbonate Reservoir Rocks," in *The Petroleum System—From Source to Trap*, ed. L. B. Magoon and W. G. Dow, American Association of Petroleum Geologists Memoir 60 [Tulsa, OK: American Association of Petroleum Geologists, 1994], 148)

issue—ideas that depend on continuity through time—the answer would be yes, for discontinuous samplings would no longer suffice (why offer snapshots, when the entire movie is available at no extra charge?).

One last example. Figure 12.3 provides a simple, vertically oriented chart, showing average measured porosity for petroleum-bearing rocks originally deposited in different settings. The data are averaged from a wide range of samplings and are thus quite generalized; the graph is meant to show only very large-scale comparisons. As a result, it probably doesn't need either the detail given in the vertical scale or the related horizontal ruling. However, we do need to know what average porosity is measured in; what do the numbers along the *y*-axis represent?

The bars are generous in width, but are not separated (as they should be), and the patterns are confusing—are we to assume some relation between those similar patterns (e.g., shoals and slope)? Confusion is avoided if even a little space is added between each bar and a single pattern used.

Note, too, the identifying text along the horizontal axis: again, are we being given clues that "Reefs" (written differently) and "Slope" (in larger type) merit particular attention? Uncertainty here can be averted in several ways, for example, by using different patterns for each bar and constructing a legend; by placing the respective label *above* each bar (which are at different heights) or even *within* it; or, again, separating the bars a little and making the type size a bit smaller and consistent.

GRAPHS

Nonbar graphs come in even more numerous varieties. With graphs you can show trends, correlations, and frequency distributions (various line graphs); rates of change (semilogarithmic graphs); changes in relative difference (area graphs); patterns among discrete random variables; and much more. The essence of a graph, however, in a majority of cases, is to show continuous relationships: data exist in some type of continuum defined by dependence between variables.

Figure 12.4 presents a simple and wholly effective semilogarithmic graph, where time is the independent variable and is given in logarithmic scale. Nothing fancy is needed here; the graph compares changes in external gill length of tadpoles hatched at three different ages (5, 6, and 7 days). The text is clean, minimal, consistent. The axes are neatly labeled. The lines are clear and easily distinguished, and the data points are marked. Note that the logarithmic scale (*x*-axis) shows actual values (0.1, 1, 10, 100, etc.),

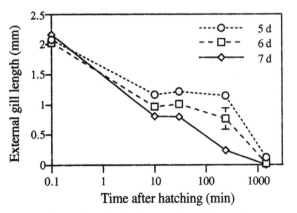

12.4   Example of a semilogarithmic graph (from K. M. Warkentin, "Environmental and Developmental Effects on External Gill Loss in the Red-Eyed Tree Frog," *Agalycnis callidryas: Physiological and Biochemical Zoology* 73, no. 5 [2000]: 559; used with permission)

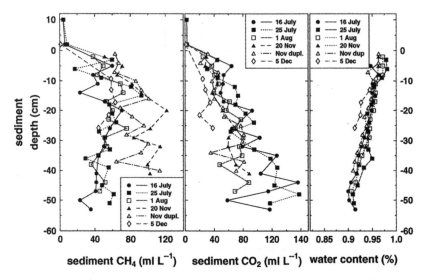

12.5   Example of a line graph (from D. D. Adams and M. Naguib, "Carbon Gas Cycling in the Sediments of Plußsee, a Northern German Eutrophic Lake, and 16 Nearby Water Bodies of Schleswig-Holstein," in "Advances in Limnology," special issue, *Archives of Hydrobiology* 54 [1999]: 94; used with permission of Schweizerbart Publishers, www.schweizerbart.de)

not the logarithms (−1, 0, 1, 2, etc.). For much larger or smaller numbers, it is common to write in powers of 10, that is, 10^5, 10^6, 10^7, or 10^{-4}, 10^{-5}, 10^{-6}, and so on.

Compare this graph with figure 12.5, which includes three related plots, showing the concentrations of methane, carbon dioxide, and water in sediments of increasing depth below the sediment-water interface (marked as 0 on the vertical axis). Each graph carries six lines—six data sets—indicated by well-chosen standard symbols. There is some question as to whether the data for methane and carbon dioxide can be meaningfully distinguished and interpreted beyond general trends (which begs the question of why individual data lines are needed). In the graph for water content, meanwhile, this is less important, because of the tight clustering. How might the graph be improved? Here is a case where using color would make eminent sense. Possibly, the plots for methane and carbon dioxide would gain a bit of clarity (and meaning) if the intervals on the horizontal scale were widened. Finally, only one legend is really needed (it's the same for all three graphs).

Deciding how many lines to plot, how many correlations to reveal, is obviously very important and can require experimentation. As figure 12.6 shows, it is sometimes possible to include many lines in an effective man-

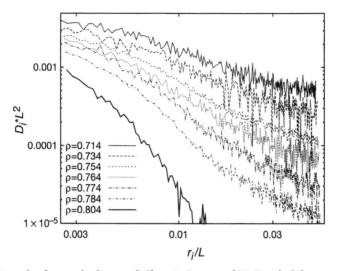

12.6   Example of a complex line graph (from L. Santen and W. Krauth, "Absence of Thermodynamic Phase Transition in a Model Glass Former," *Nature* 405, no. 6786 [2000]: 550; copyright 2000 *Nature*; used with permission)

ner. Notice here how well-chosen the line patterns are in terms of the eye's ability to differentiate—an aspect that adds both clarity and visual appeal. At the same time, however, this is possible only because the lines do not cross each other very much. Were they to do so, it would be necessary to break the data out into two or three separate graphs. Visually speaking, moreover, not all lines are created equal: note how the thicker, solid plots draw attention and imply importance, while the thinner, more broken ones carry much less psychological weight. Perhaps the only way to improve this example is to make the y-axis more consistent in interval labels, that is, 10^{-5} (we don't need the "$1 \times$"), 10^{-4}, and 10^{-3}.

Another type of line graph, sometimes called an area graph, is given in figure 12.7. This type of data display is used to show both progressive change and comparisons between different data fields. Note how important it is to use different and easily distinguished patterns within each data field. Color is not necessary here, but could be a help if a larger number of fields were plotted. Small improvements might be made: the axes might be better labeled ("MCFG" might be written out along the y-axis, "Million Cubic Feet of Gas"); we might want to extend the data fields through dashes (projected) to the right margin of the graph. On the whole, however, this is an informative and well-done figure.

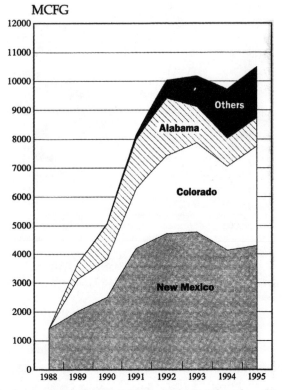

12.7   Example of an area graph (from D. K. Murray and S. D. Schwochow, "Coalbed Gas Development in the Rockies—Analogues for the World," in *Innovative Applications of Petroleum Technology*, ed. E. B. Coalson, J. C. Osmond, and E. T. Williams [Boulder, CO: Rocky Mountain Association of Geologists, 1997], 32; used with permission)

Illustrations

Figures 12.8 and 12.9 show maps with different sorts of problems. Figure 12.8 is meant to compare precipitation levels associated with the 1997–1998 El Niño phenomenon, but the maps are far too small to read and interpret on any intelligent basis. In such cases, the editor of the journal and the authors of the article need to make a decision that favors their readers, not the data alone. If no more space could be allotted these images than is shown, they should have been deleted or else one map selected and shown at larger size. As it is, they provide more in the way of frustration than enlightenment.

This brings up an essential point. Authors very often use color to try to

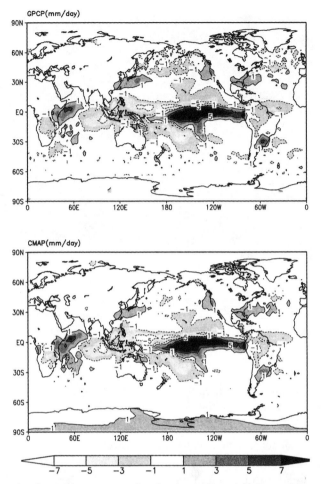

12.8   Example of maps that are too reduced to be easily read (from A. Guber, X. Su, M. Kanamitsu, and J. Schemm, "The Comparison of Two Merged Rain Gauge-Satellite Precipitation Datasets," *Bulletin of the American Meteorological Society* 81, no. 11 [2000]: 2641; copyright American Meteorological Society; used with permission)

overcome the problem just described. In fact, since color visuals are now nearly ubiquitous in science, authors tend to assume that they are able to minimize difficulties in legibility for shrunken or even miniaturized images. They don't. Imagine figure 12.8 in color, with gradations from red (darkest areas) to orange, then yellow. Would this help? A little, perhaps; the image would be more attractive but not much more informative. This is because the smaller the image, the more generalized, more simplified, and

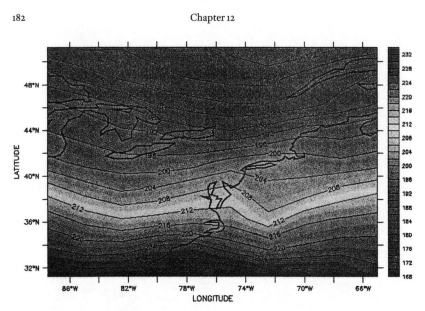

12.9   Example of a map that is difficult to read due to loss of color (from T. D. Bess, A. B. Carlson, C. Mackey, F. M. Denn, A. Wilber, and N. Ritchey, "World Wide Web Access to Radiation Datasets for Environmental and Climate Change Studies," *Bulletin of the American Meteorological Society* 81, no. 11 [2000]: 2649; copyright American Meteorological Society; used with permission).

less interesting it will be. If you or a coauthor has made the effort to map such data in detail and derive good meaning from it, you are wasting that work when you reduce it to the borders of legibility.

Figure 12.9, meanwhile, offers a related reminder. Images drafted in color for talks or posters do not always translate well for published articles when rendered into black and white or grayscale. Here, the data would be illegible if the contours weren't numerically labeled; certainly the scale on the right is of little help. The figure would be much clearer if it lacked shading altogether and merely offered its contours against a white background. Were color to be restored to this map, however, its message would be fully recovered but *only if* all numbers were legible. If not, if through laziness or oversight they appeared blurred or tiny, both your time and the reader's time would have been wasted.

Which brings us to the example of figure 12.10. Let us admit, up front, that digital presentations are a major advance for scientists. Indeed, they are far more simple, efficient, and (even, sometimes) fun to create than the old, laborious hand-drawn figures and typed-out tables of former epochs. But presentation figures often need to be modified for inclusion in a pub-

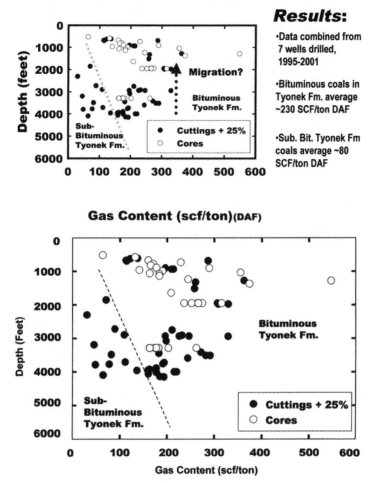

12.10   Example of a computer-generated graphic. *Top*, unedited from use in oral presentation; *bottom*, properly edited for print.

lished paper. A graphic like the top panel of figure 12.10, with its verbiage surrounding and inhabiting the data, plays fairly well up on the big screen, but looks clumsy and amateurish in a journal article. Much of the writing here is interpretive and belongs in the main body of the text. To appear in a high-quality journal, the graphic would need to be revised as shown in the bottom panel of figure 12.10: title removed, interpretive text gone, axis labels reduced in size, lines thinned.

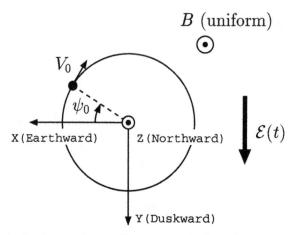

12.11   Example of a schematic diagram (from M. Nosé, S. Ohtani, A. T. Y. Lui, S. P. Christon, R. W. McEntire, D. J. Williams, T. Mukai, Y. Saito, and K. Yumoto, "Change of Energetic Ion Composition in the Plasma Sheet during Substorms," *Journal of Geophysical Research* 105, no. A10 [2000]: 23283; copyright 2000 American Geophysical Union; reproduced with permission).

Schematic diagrams are quite common in science, and, as their name implies, are usually most effective when kept as simple as possible. Figure 12.11 is an illustration of a dipolarization model that shows this very well. The image employs basic shapes and visual elements (circles, dots, arrows). It makes good use of space, being neither crowded nor too open. It includes a minimum of text; in fact, the use of words on the diagram is a good thing, since it gives us valuable orientation and prevents the figure as a whole from becoming a mosaic of single-letter symbols.

Another good example is figure 12.12, which depicts how a particular type of bone responds to an applied stress. In this case, the diagram shows a cyclic progression of states, involving a temporary weakening of bone mass (through the formation of a cavity and resulting increase in strain) and its repair. A reader can follow the process easily, from left to right. Again, there is minimal text, just enough to identify crucial elements (all abbreviations are explained in the caption). Color would be an effective addition but is not necessary here, since shading can accommodate everything shown. Is the figure perfect? No such phenomenon exists. Some scientists would find value in numbering each of the stages and using these numbers to explain the process in the caption or main text; this would be efficient and effective and would not add overly to the information shown.

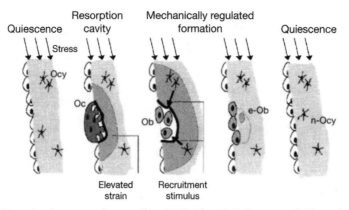

12.12 Example of a process diagram (from R. Huiskes, R. Ruimerman, G. H. van Lenthe, and J. D. Janssen, "Effects of Mechanical Forces on Maintenance and Adaptation of Form in Trabecular Bone," *Nature* 405, no. 6787 [2000]: 705; copyright 2000 *Nature*; used with permission)

Others might provide a legend explaining each of the shaded areas, as well as abbreviations.

Combination diagrams, in which two or more types of illustrations are grouped together, have become quite common in many scientific publications. This has led to many fertile blendings of visual information. In figure 12.13, images and spectral graphs are nicely juxtaposed to illustrate behavior of alternating ferromagnetic and antiferromagnetic layers within a given substance. This figure contains a large amount of information, but presents it clearly and even elegantly. Note, for example, how each half of the total is visually balanced, with the legend inserted neatly into the upper right portion of the respective graph (which would otherwise show a lot of white space). Small arrows are used to tie portions of each image to its respective position on the spectral graph. Though the overall size of the figure is small, the text and numbers are large enough and in an appropriate font to be easily legible.

Another example is given in figure 12.14. Here a specimen drawing is combined with a histogram. The drawing indicates the few key anatomical features of the cricket relevant to the experiments, with the most prominent of these (HW) shown in capital letters to emphasize their central importance. The histogram, meanwhile, plots two types of escape behavior for four different kinds of stimuli (plus a control group), each of which was applied to the HW (hindwing tip). At first, everything looks to be in good order. But when we examine the details of the figure, a few problems

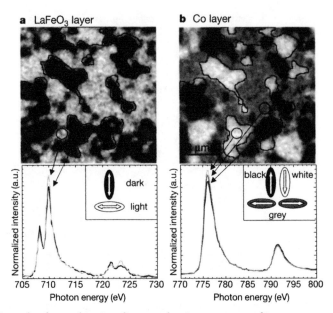

12.13 Example of a combination diagram, showing two types of images successfully integrated (from F. Nolting, A. Scholl, J. Stöhr, J. W. Seo, J. Fompeyrine, H. Siegwart, J.-P. Locquet, et al., "Direct Observation of the Alignment of Ferromagnetic Spins by Antiferromagnetic Spins," *Nature* 405, no. 6788 [2000]:767; copyright 2000 *Nature*; used with permission)

emerge. Notice that there are two sets of *a* and *b* images, one in uppercase, the other in lowercase, plus a *c* designation on the upper specimen drawing. Probably the lowercase *a* and *b* could be replaced with numbers or deleted altogether. It would also help, to avoid any confusion, to place the legend (black box, open box) at the bottom. There is also enough space in the figure to write out the words for the anatomical features indicated on the drawing, none of which is very long (foreleg, midleg, hindleg, circus, hindwing). This would reduce the burdens on the caption to list and explain a large number of symbols, a consideration worthy to be granted the reader.

As a final example, I turn to an illustration that reveals what is now a fairly standard level of visual diversity and complexity in biomedical imagery. Figure 12.15, from the premier journal *Cell*, shows seven images of varying types merged into a single visual conglomerate. These images include a schematic drawing, two line graphs, a scattergram, a histogram, and two tunneling electron micrographs. The caption, not surprisingly, reads

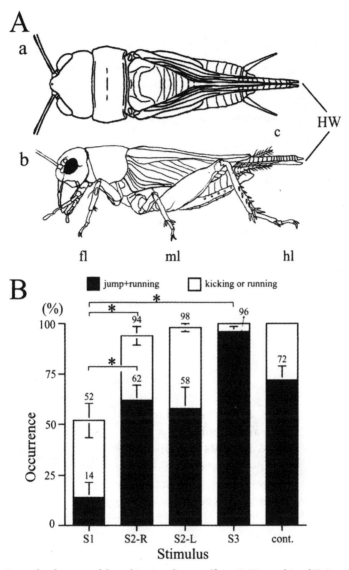

A

a

HW

b

c

fl   ml   hl

B

■ jump+running   □ kicking or running

(%)

Occurrence

*

*

*

94

98

96

52

62

58

72

14

100

75

50

25

0

S1   S2-R   S2-L   S3   cont.

Stimulus

12.14 Example of a successful combination diagram (from T. Hiraguchi and T. Yamaguchi, "Escape Behavior in Response to Mechanical Stimulation of Hindwing in Cricket, *Gryllus bimaculatus*," *Journal of Insect Physiology* 46 [2000]: 1332; copyright 2000 Elsevier Science; used with permission)

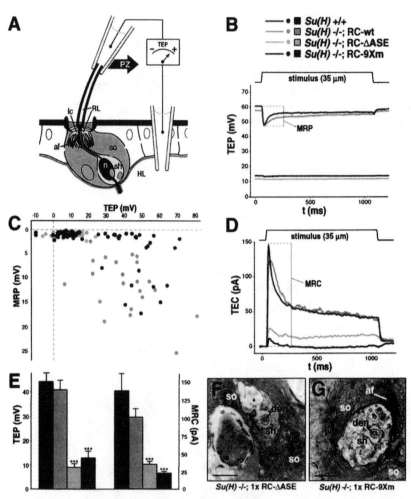

12.15   Example of a complex combination diagram, involving a wide range of image types (from S. Barolo, R. G. Walker, A. D. Polyanovsky, G. Freschi, T. Keil, and J. W. Posakony, "A Notch-Independent Activity of Suppressor of Hairless Is Required for Normal Mechano-receptor Physiology," *Cell* 103, no. 6 [2000]: 964)

like a headline: "*Su(H)* Auto-Activation is Required for Normal Mecha-noreceptor Function," with the following explanation of each segment (A–G) requiring nearly half a page. A few decades ago, editors would have rejected such an image. Now it is a *modus illustratis* in certain branches of science. It allows far more data to be shown, creating something like a miniarticle within the larger paper or report. Obviously this is an approach that can easily be taken too far; visual chaos and confusion are the risk. Yet readers have adapted to such displays, and writers must learn how to produce them in acceptable forms. What they suggest, after all, is scientific imagery pushing toward a certain limit. Beyond it? Animated or even video imagery, perhaps, with narrated captions.

Referring to Illustrations in Your Text

Part of the craft (and sometimes art) of good scientific writing is knowing how to integrate graphical materials. Visually, illustrations are embedded in the text, like boulders in a stream, but you want them to be part of the flow as well. Perhaps a better metaphor is the quilt: graphics need to be sewn or woven into the larger pattern of meanings. The reader should be told their significance, why they are there and what they show, and this needs to be done at certain points in the narrative, and in certain ways. How and where you refer to your visuals can make a subtle but very important impact on the reader's experience of your document.

There are two basic ways to refer to figures in the text. One is indirect, with the reference placed in parentheses, for example, (fig. 1.2); this is the most common form. The other approach is direct and makes the graphic a subject of discussion, for example, "Figure 1.2 shows . . ." Deciding which form is best often means choosing whether you want your reader to glance over a particular figure, or study it in extensive detail.

No hard-and-fast rules exist to guide you in your decision. But some commonsense precepts can be applied. If a particular graphic is used to demonstrate or establish an important finding, or to suggest a central interpretation or conclusion, it is appropriate (though not necessary) to mention it directly, making it the subject of one or more sentences, even a paragraph. On the other hand, if you include a visual to illustrate a point made in the text, to give an example, or to sketch a piece of apparatus, then indirect reference is probably sufficient.

Among our examples, a few figures that might deserve direct discussion are 12.2, 12.11, and 12.12. Each shows information that is central to the narrative, representing a major result of the experiments performed. Indirect reference, meanwhile, seems more appropriate to graphics like 12.3, 12.7, and 12.10, which provide generalized data or schematic representations. In the end, such decisions can often be subjective. If you have doubts about how to handle a particular graphic, look at your models or check the recent literature to see how other authors have dealt with similar images.

After you've made your choice as to direct or indirect reference, where should you place it? In what part of a sentence or paragraph? Too many authors, when it comes to indirect mention (in parentheses), jump the gun and insert the reference before the reader is ready. Examples abound in the literature. Here's one:

> The conventional structure of a bipolar transistor (Fig. 2A) requires three distinct material types: for example, a highly doped n-type layer . . . , a p-type base, and an n-type collector.[1]

Note that the eye is told to abandon the sentence even before enough information is given for the figure to make sense. How are the authors using the figure? As written, they are emphasizing "conventional structure." But the actual message of the sentence and paragraph, and the figure too, focuses on the three parts of this structure, not its conventionality. Thus, it would be much better to place the figure reference either after the word "types" or, best of all, at the end of the sentence.

Another example relates to figure 12.7:

> Proved [natural gas] reserves estimated by the U.S. Energy Information Administration have risen from 1.4 BCFG in 1988 to 10.5 BCFG in 1995 (Fig. 2 [our fig. 12.7]), more than 70% of which occurs in Colorado and New Mexico.[2]

Again, the reference belongs at the end of the sentence, after the pertinent information has been given. This cues the reader that *all* of the values mentioned are displayed on the figure, rather than reserves figures only. Note, too, that the reserves values given in the text are in BCFG (billion cubic feet of gas), rather than in thousand MCFG (million cubic feet of gas), as shown in the figure. This sort of discrepancy should be avoided. Using the same units in text and illustrations is needed to integrate the two.

Similarly, try to use phrases and terms in your captions that are picked up in the text. Much could be said (and has been) about how to write proper captions, what their length should be, whether to use full sentences or phrases, and so on. Standards here, as in so many other aspects, tend

---

1.   Y. Cui and C. M. Lieber, "Functional Nanoscale Electronic Devices Assembled Using Silicon Nanowire Building Blocks, *Science* 291, no. 5505 (2001): 852.

2.   Murray and Schwochow, "Coalbed Gas Development," 32.

to change between fields and between journals. Make sure you're aware of any restrictions or rules given for your target journal or publisher. As always, where there is any doubt, use your models as guides.

A Final Point

Scientific illustration, as it exists today in wondrous plenitude, can never be covered adequately by a single chapter such as this. Indeed, an abundance of volumes have been written on the topic, even on such seemingly humble forms as the chart or graph.[3] I've tried to cover some of the more obvious and necessary aspects of creating good graphics, and of evaluating those of others.

Beginning authors, once familiar with the literature, may feel (and even be thankful) that the types of illustrations available to them for their own writing appear to comprise a set of fixed formulae. Closer inspection will show that this is rarely, if ever, true. For many early-career scientists, it is helpful to stick to the most common patterns. But experienced scientist-authors understand that illustrations are actually a flexible means of expression, and can be adapted, albeit conservatively, to individual cases. Close study of premier journals will prove this: different authors add modifications, sometimes small, sometimes significant, to what are otherwise standard graphical templates, and they do this in order to make their message a bit more efficient and elegant. Here, too, just as with writing itself, one can ultimately choose between functional and creative approaches.

---

3. A few sturdy works in this field include Robert S. Wolff and Larry Yeager's *Visualization of Natural Phenomena* (1993), Robert R. H. Anholt's *Dazzle 'Em with Style* (1994), Mary Helen Briscoe's *Preparing Scientific Illustrations* (1996), and, of course, the several well-known books by Edward R. Tufte: *The Visual Display of Quantitative Information* (1983), *Envisioning Information* (1990), and *Visual Explanations* (1997).

*

13. ORAL PRESENTATIONS:
A FEW WORDS

*Before I speak, I have something important
to say.*—GROUCHO MARX

The Spoken Word in Science

Talking is one of the things that human beings do best. This is true, even of scientists. While no small difference separates a hallway chat from a formal talk, both involve sharing knowledge through the spoken word, the most basic form of communication. An oral presentation is an opportunity for you to be the only local speaker in science's unending conversation. One side of the conversation falls silent for a few moments, with interest and attention, to hear about you and your work.

It is not generally known that the beginnings of modern science largely derive from oral sources. The very first journals, which began publishing in the 17th century—the *Journal des Sçavans* in Paris and the *Transactions of the Royal Society of London*—were composed mainly of transcribed lectures. These lectures were given before each society, the Académie des Sciences in France and the Royal Society of London in England, usually at the rate of one per week and in a seminar format. Members were free to ask questions, offer criticisms, and discuss particular points. The original speech would then be revised and submitted for publication in an upcoming issue. Though each lecture was written out beforehand, it is likely that much ad-libbing went on during actual delivery. Writing and speaking were inseparable, like color and taste in an excellent wine. The tradition was carried on well into the 19th century, when the great rise of journal publication began.

Today, oral communication is integrated into scientific work at every level and in every sector—think of classroom teaching, seminars, office banter, parties, departmental or team meetings, presentations to management, symposia. We are all, in a sense, professional talkers in science. Consider: during a single week, you are likely to give any number of little "speeches" about your work, whether to a colleague, advisor, student, administrator, manager, someone in another discipline, a family member, or the bathroom mirror.

Oral presentations take this very human reality and give it a more official, ceremonial cast. You may feel, sometimes, in a moment or two of stage fright, that such talks are the penance to be paid for a poor career choice. But this is only momentary anxiety speaking. Formal talks are the special opportunity we give ourselves to discuss our private work in a public voice, to make the labor of our minds and hearts communal.

Essential Attitudes

Listening and reading are very different activities when it comes to absorbing knowledge. If written material presents a chosen residue, the spoken word is a communicational volatile: once offered, it vaporizes. Writing is static, permanent, and allows for repeated study. Speaking is dynamic, transient, and reliant on human interaction. As a result, the "stories" we tell by voice need to be different from those we narrate in an article or report. And part of this difference resides in the fact that you, the speaker, put a human face on things, a face hopefully of enthusiasm and interest.

The professional talk is a kind of lecture. For a few moments, listeners are willingly transformed into learners, and you become their instructor. This does not mean they are available for condescension—they are your equals in every sense except with regard to the specific knowledge you are about to impart.

Is your presentation a type of performance? Yes, of course—but not merely this. You don't have to deliver a presidential address to do your work well. But it will help your message enormously if you give it *some* personality. Your audience is already captured, so to speak; presenting to them with a degree of enthusiasm is being considerate and professional. Dull and monotonic lectures are a type of insult to those who are spending money and time (away from their own research, their families) to come listen. If your voice and manner display interest, your audience will too.

Why do I emphasize these realities? Because they can help give you direction. If you are a teacher as well as a researcher, think of an oral presentation as a natural extension of your classroom efforts. If not, consider it an opportunity to practice teaching skills that might serve you excellently in the future (e.g., if you decide to make a career change).

Here are some points to consider:

- Begin to *prepare* your talk *well in advance*. Outline a basic structure for it (use headings), consider alternatives; think about visuals. Take notes on other talks you see that you consider especially good.
- Design and write out your talk in a manner you *feel comfortable* with, whether this involves using an outline, note cards, storyboarding (laying out visuals in order, with a bit of text for each to guide what you are to say). Experiment, if necessary, to see what method you can speak from the most easily.
- *Be* sure you are *completely familiar* with every aspect of your visuals, including all their details, as if you created them yourself. You should be able to explain just about every aspect of them (Why this scale? What about those data points upper right? What river is that?). This is an essential part of having command over your material.
- *Practice your talk*, when it is ready. This is paramount. Practice again and again (at least 10 times, for good measure). Do this by yourself and in front of colleagues, until it becomes fluid. Practice breeds confidence and a sense of control over the material. There is no substitute for this. Think of what it takes to perform a piece of music in front of an audience.
- *Practice portions* of your talk *that are difficult for you*, less smooth and confident than the rest. You can simply do this in your head as well as out loud if you like, but you need to do it until your confidence is strong enough that you won't stumble or pause overly long to gather your thoughts.

When you practice, pace yourself. A good rate of speed is about 90–120 words per minute (average 100). This means for a 20-minute talk, about 2,000 words; for a 30-minute talk, about 2,800 words (providing for pauses, etc.); and for an hour-long talk, about 5,500 words. These rates are significantly slower than conversational speech, and they need to be.

Measured speech allows the audience to keep up with you, and it gives you the appearance of control and fluidity.

Stage Fright: A Universal

Please don't doubt, even for a moment, that all of us are subject to stage fright. This is natural no matter how many times we stand before the lectern, settle our notes, and gaze out into a sea (or pond) of faces. Speaking is an intense experience; one feels exposed to the ferocious scrutiny of strangers.

Yet, as veteran presenters can tell you, your listeners are far more likely to sit before you in sympathy. Many or most have been there, and know what it feels like. If asked, they would readily admit appreciation of your efforts to share your work. It is thus with these perceptions—which, in effect, validate the very reasons for giving oral presentations—that you yourself should begin.

Experts on public speaking will tell you that stage fright has a positive side. It is not merely fear, but also excitement. We wouldn't feel nervous if we didn't care about our performance or the content of what we're about to offer. This suggests that you might try to aim your nervous energy in positive directions. Breathe slowly; think of your speech as an act of intelligence and bravery (which it is). Visualize yourself doing well, moving through the material easily. Hear your voice saying the opening sentences of your talk, then going on from there.

The spoken word is the music of knowledge: momentary, personal, expressive. No matter how scripted it might be, how calm and replete with specialized terms or tones, it emerges from you, the individual speaker.

Structure and Flow

When you design your talk, try to establish a fluid logic between your main points. It can help to write out the primary topics you think you might want to cover, or lay out the visuals you may want to include, then search for an order among them. Don't be afraid to do this at an early stage. Try to experiment: include more topics or visuals than you might need, move pieces around, take some out, see if a particular sequence appears.

In doing this, think of someone in the audience who will be taking notes on what you say. He or she should be able to put down your major points, in order, without difficulty. Your presentation, that is, should include discussion of its own structure. It should cover the following:

- Why did you do this work? (emphasis on "this")
- How did you do it? (the tools, techniques, approaches you used)
- What were your findings? (your results)
- What you have concluded from it? (what it all means)

If you are able to cover these areas, in order, within the time allotted, and without causing your listeners to go unconscious, you have probably done a good job.

How to begin? In most effective presentations, a speaker will start with a word graphic showing his or her name, affiliation, and the title of the talk. This may seem unnecessary, given that the talk has been listed and an abstract probably made available. But human reality requests certain considerations. During the first few minutes, the audience will be adjusting from the previous talk, settling into their seats, still leaving or arriving, sipping coffee. A moment or two of introduction is therefore needed; it will also help get you, the speaker, acclimated and focused.

Next, it is a good idea to show the audience how your talk is organized. This can be done using a second word graphic covering an outline of the major topics. Introduce this material with a phrase such as "today we will be talking about" or "in this presentation, I'd like to cover the following." Your graphic, meanwhile, can provide a list of brief phrases devoted to each of the above questions (why you did this work, etc.), or a general overview (three or four areas to be discussed). In all cases, it works very well to present these points at the beginning and then to return to each one, sequentially, later in the talk, when you treat it in detail.

Think of your talk, therefore, as an hourglass divided into sections. You begin speaking in general terms (e.g., your research topic and its importance), move to greater detail (methods, findings), and end more broadly (what your work means, its implications). Within this overall shape, you'll have different parts that may or may not correspond with these larger stages. For example, you may want to devote a section each to overview, methods, initial findings, final results, conclusions, and future work.

You might begin each new section by repeating your outline and highlighting the portion that will next be discussed. Cue these transitions verbally, too, with such phrases as "let's turn now to" or "having looked at . . . , we can now examine." This kind of segue, repeating and newly highlighting part of the outline, isn't really appropriate for short talks of 10–15 minutes, but can be very helpful to the audience in longer presentations.

Finish up with a review of your major points, followed by a set of final conclusions. Keep these as simple and direct as you can, as they will be the last things your audience will see and hear. Between your opening and closing, you want to move your listeners through your story as smoothly as you can. Try to order your visuals so that the audience can easily follow your logic from one to the next. Think of including a few "relief" images, that is, those that are more purely visual (photographs, models, maps) and might give your listeners a break from a series of text- or number-heavy graphics.

Finally, don't think of the time limit (10 or 20 minutes, usually) as a burden, but instead as a kind of liberation. Why? Consider the difference between writing a full-scale research paper and an abstract-like summary—which is easier? Would it really be less work to produce an hour-long speech on your topic? Working within the given time limit gives you the freedom to hit the high points and do no more. It releases you from the bond of having to prove every generalization or important statement; you need support only those you feel are most central to your story. A talk, moreover, gives you the chance to tell your tale according to the simplest, most Aristotelian of formulas, with a beginning, middle, and end, and a flow that follows the basic scheme of "I did this because . . . using these tools . . . and having found these results . . . made these conclusions."

Using Visuals: Realities and Techniques for Electronic Display

In today's world, electronic forms of display utterly dominate at scientific meetings. No matter what else may be said about it, the use of presentation software marks a tremendous advance over the past, when Paleozoic forms like transparencies and photographic slides were common. These were options whose demands on time and money led many (yours truly included) to hope for their eventual extinction. Well, it has come. The new digital era has fully arrived. Not, however, without a mix of sunlight and clouds.

First, let me say that visuals in an oral presentation are far more than mere props or aids. They are very often the main bearers of the message. In more than a few cases, they *are* the message, as when images of specific phenomena (an atomic structure, a landform on Mars) are shown. This means that all visuals must be clean, sharp, and easy to read; if they aren't, your talk will fail at each point where this isn't the case. It will fail, because your true purpose, which your audience understands, is not to *show* but to *communicate*, not to fill time but to convey meaning. This may require you to make changes on an original image, resize it, crop it, add arrows or circles, whatever is needed to make it fully visible and articulate *without* altering the substance in any way.

Presentation programs allow you to do many things to make your images more eloquent and more interesting. Indeed, different programs offer different capabilities along these lines. Even as these programs multiply in number and advance in options, however, far too many talks continue to simply show static visuals. This can be described as vestigial, a behavior inherited from a former era and species. Some images, to be sure, are complete and need nothing more. But a great many could do with a little life and some with more. It is a simple matter, for example, to highlight different parts of a visual in succession as you make a series of points, one at a time. Alternatively, you can have different portions appear one after the other, so that they build a final visual, or you can remove pieces one by one to focus in on a key part. Similarly, having key points appear one by one on a slide, as you discuss them, is easy and can be effective if used strategically and not too often. Some programs give you the ability to magnify portions of an image or to rotate it so that certain features can be better seen. Most allow you to highlight (and de-highlight, which is necessary) areas on a visual or key words to help you be emphatic at a certain point in a talk.

These are very simple animations and are easy to learn, but they can do a lot to engage a viewer. Beyond them, you can take a talk to an altogether higher level. Presentation programs now provide the means to embed video or audio, plus real-time animations that are highly instructive. Using these capabilities is not considered flashy or flamboyant if the material is content-rich or necessary. In my field (geoscience), it is no longer rare to see video clips providing examples of geologic process, aerial views of landforms and structures, models of deep-earth activity, and more. Though not standard, by any means, the occasional appearance of such visuals may well become more common over time. Each field, of course, will have its own level of acceptance and use for these options, and scientists will need to decide for themselves whether they wish to utilize them. In my own experience, when well chosen, these higher-level displays are both enjoyed and appreciated and add greatly to the overall impact of a talk.

The variety of programs now available for scientists, meantime, is larger than ever. That so many continue to use Microsoft PowerPoint is far more a matter of historical momentum than available choice. The truth of the matter is that it is part of your responsibility as a speaker to become familiar with at least some of these programs and whether they can help you communicate more effectively. In other words, it's worth trying them out, experimenting with them (not in front of an audience), to see what they can do for you. Some programs, like Apple's Keynote, can be used for any discipline. This is also true for a number of online, web-only programs such as Zoho Show, Haiku Deck, Google Slides, and Prezi. Others, such as ScienceSlides (VisiScience), are specific to certain fields, such as biomedicine. The goal is not to become overwhelmed, but to experiment, try things out, and see what works best. It is inevitable that more products of this kind will become available (the names I have mentioned may well no longer exist in 5 or 10 years). As always, it is worth taking note of those presentations you yourself find excellent and learning from them.

There are some dangers to be aware of too. Such visual techniques can be overused (not *every* slide needs them) or misused (some images don't need them) or even abused (employing the fanciest glows and glimmers). Trying to be clever or flashy will instead make you look immature and self-indulgent. Still worse, it will trivialize your material. You cross the borderland between success and failure whenever such techniques draw attention to themselves, thus away from the science. Use them somewhat sparingly, therefore, and always in mind of the real message. If you are unsure about anything, show it to a colleague or (if no one is available) try to view it from the viewpoint of a speaker whose talk you admired: would she or he use something like this?

Which brings us to a larger question about medium-sized matters. There have been claims, even by well-known people, that the use of presentation software "makes you stupid." Having used such programs for well over a decade now, on a weekly basis, I find I am still literate and can tie my own shoes, but I also understand what these critics mean and that they have a point. The ease of using a program like Microsoft PowerPoint can easily encourage laziness in how it's used. Examples, in fact, are rife. Here are the worst, and unfortunately the most common, of them:

- Dumping far too much information on slides
  - Multiple diagrams (Visual overload!)
  - Overly complex images (These must be *studied* to be understood.)

- ○ Dozens and dozens of words (Do I *read* or *listen?*)
- ○ Too many slides with no visuals at all (Why am I *looking* at this?)
- Killing the audience with bullet points
  - ○ Overuse (Slide after slide after slide after . . . this is beyond boring.)
  - ○ Too many on one slide (The maximum is approximately five, but four is better; if more, the mind goes numb.)
  - ○ Poor or confused phrasing (Equals lost attention; what does *that* mean?)
  - ○ Thoughtless list of "main points" (Creates scatter, not summary; there should be a connecting logic in which one point flows into the next.)

These problems add up to a confession. They say, quite clearly: you are a poor speaker and you know it, but you don't care enough, whether about communicating your research or about the audience and their experience, to take the time to improve things. Little of this may be true, but it is the message you impart.

It can help greatly to consider your own response in meetings you've attended (suffered through). Better still, find or call up some presentations of other researchers in your field and go through them, slide by slide, with a critical eye. Remember that in most talks, each slide will be up for roughly one minute or less (if there are speaking notes, you can gauge this better). What works and what doesn't? What would you do to make any single slide better? What seems well done, and how can you use it for your own talks?

One of the above problems can be simply demonstrated. Here is an example of a bad set of bullet points from a discussion about landslides:

Frequency of Occurrence

- Weak material means slopes are unstable and subject to failure
- Major seasonal meteorological events that raise opportunities for failure
- Oversteepened slopes
- Glacial deposits
- Stream erosion into hillsides
- Human development often happens where landslides have occurred in the past
- Lack of prediction capability

The first thing we might notice about this slide is that its title and contents don't match. A far better title for most of the points would simply be "Why Do Landslides Occur?" Note that such a title, using ordinary language, is perfectly fine and even excellent in a professional research presentation; it signals that what follows will also be simple and clear (so it should be). The second aspect to fix our attention is that there is no consistency in wording: some points are two-word labels, others are brief phrases, and still others sentences. Some are nominative (nouns), some descriptive ("Lack of prediction capability . . ."), and one is even narrative ("Human development . . ."). At least one point, the last one, doesn't belong here at all.

Such are the signs of info-dumping. The speaker has written down what seems relevant, without much logic and without thinking very much about *how* things should be expressed. The speaker then moved on to the next slide and never returned, not really, since she or he didn't take the time to read carefully and revise what had been spilled on the slide. The speaker may think: "I'll make this clear enough when I speak." But we know, too well, that this is wishful thinking. The Muse is not so readily agreeable when discipline is lacking. How might we fix what is broken and incomplete? Here is a suggestion:

Why Do Landslides Occur?
- Steep slopes of loose material
- Intense seasonal rainfall
- Stream erosion into hillsides
- Human development

All of these lead to slope instability and failure.

In this case, I have decided on short, simple phrases. I've also given them a particular order: first, given conditions, followed by processes acting on these conditions, leading to creation of instability and, finally, failure. It provides a straightforward, deductive flow. A couple of points have been removed, namely "glacial deposits" and "lack of prediction capability." The first is unnecessary, since it merely gives one example of "weak material," while the latter, as noted earlier, belongs elsewhere. There is now enough room on the slide to add an image that illustrates some of the points. Depending on the rest of the talk, you may or may not wish to do this. It is, however, an option.

A final observation is worth stating. You have undoubtedly gathered that I disagree with those who feel presentation programs are, by nature, demonic or "stupidifying," forcing us to make the kind of cognitive errors noted above (and worse). With all due respect to the perception, it is what Freud once called the error of mistaken transference. This is because the widespread phenomenon of static, poorly organized, unclear visuals can be said to reflect something fundamental: how difficult it is to communicate science effectively in this way. We might compare it to teaching: how many courses do any of us recall where the instructor and the images he or she used were eloquent enough to rivet our attention, animate our interest, and bring the material vigorously to life? Classes of such caliber are rare as black opal and far more valuable.

Yet the answer to this sobering question has a positive side too. Just as there do exist excellent teachers, there are good, admirable, and even brilliant speakers in science, whose visuals are the equal to a dynamic personality. Their talks are the ones we need to pay special attention to and learn from. When you find yourself sitting before such a speaker, focus on what is shown (take notes, if it makes sense to do so) and how it is discussed. If not prohibited, use your phone to record portions of the talk, even the whole thing; otherwise request a copy of the presentation from the speaker or ask him or her whether the presentation is available online. If you feel shy or embarrassed doing this, keep in mind a simple truth: even if his or her research is much less significant than yours, it will be far more memorable (and will *seem* more significant) if the presentation is superior.

Successful speakers therefore often plan their talks on the basis of the images they intend to show. Whether you adopt this approach or not, it is important, very important, to plot out visuals (or types of visuals) at the same time that you plan out the structure and organization. A well-proven technique for doing this is to storyboard your talk. This involves the actual mapping out of a presentation, whether on paper or by presentation software. Common methods involve using one sheet of paper or slide per visual, with the graphic placed (or described) in the upper half and notes for speaking underneath. Any presentation program today allows you to do this easily. The result has the advantage of allowing you to flip through your talk, making any additions, deletions, or changes in organization immediately. It can also produce a rehearsal slide show for you, allowing you to practice and time yourself (very important) with little effort. This can really be a help, and can speed up your preparation significantly.

No less than your narrative, your images must tell a story, indeed *the* story, the one you want to get across. Therefore, choose them carefully; consider both the content they show and how interesting they are to look at. In some cases, an effective approach is to use what Michael Alley, author and professor of engineering communication, calls "assertion-evidence."[1] This is where the title of your slide states the message ("Landslides can be caused by human excavation, like building roads and houses") and the illustration provides visual evidence. Many scientists do this instinctively, particularly if they are involved in fieldwork or an area dependent on direct observation. You can avoid the less interesting textbook display—a label-like title ("Landslides result from human activity") alongside an overly perfect, completely interpreted example—by offering an uninterpreted image first, followed by a partially interpreted one, and then filling in the final details verbally while pointing them out on the image. This is more work, certainly; but it engages the audience to a higher degree, demonstrates your expertise and authority, and can even provide an opportunity for comment or humor ("*This* might not be the best place to build your retirement home").

Thinking hard and productively about your visuals means being open to new ideas, to a bit of experimentation. Even if your talks are already high quality and attract admiring comments, you can make them even better. As I've said many times in this book, advancing your communicational abilities is part of advancing your research.

A final practical point. A sine qua non for every presentation you ever do is to create multiple copies of it. You may feel this is obvious. It is, but then again, it isn't. Some venues will not let you use your own laptop. Even if they do, there may still be problems getting everything to work properly. If you use a Mac and the venue is set up for PCs, you will need an adapter, and if (for some unaccountable reason) you don't have one . . . Then, even when you do bring such an adapter, there is the wonderful experience of having your laptop crash or otherwise throw a tantrum at the appointed hour. And so on. Copy your talk onto a portable drive (USB flash drive) and have it ready. Better still, have it on a USB flash drive *and* somewhere

---

1. M. Alley, *The Craft of Scientific Presentations*, 2nd ed. (New York: Springer, 2013).

online that you can access if all else fails. Case in point: the keynote speaker who rushes to be on time, only to find that his or her laptop and USB flash drive are both in a carrying bag left in the hotel room.

Designing Visuals: Further Practical Points

As I've already said, keep track of the things you like and dislike about other scientists' visual materials in talks, what you find effective and what strikes you as confusing or ill-chosen. For example, make note of what it is that renders a particular chart or table easy to read (Well-sized type? Good contrast? Clear design?), or, on the other hand, what renders it incomprehensible (Too much information? Type too small? Poor use of color?). You might see a very nice display of routine information or an innovative arrangement of data or graphic material—note this or take a picture of it for reference. Better yet, ask the presenter for a copy—again, don't be shy; this is a great opportunity to help improve your own work, possibly to make a new contact, and to offer a compliment that may be returned to you in the future.

In some cases, the proceedings from a symposium, conference, or seminar will include printouts of slides shown during the talks. Such proceedings, therefore, can be a gold mine of visual material for you to evaluate and learn from. If at all possible, collect the best images in a file, along with the textual models you may have chosen. Review these before designing any future talks, to keep their lessons fresh.

Aside from what you will learn in this manner (a great deal!), here are some pointers to keep in mind:

- *Keep your graphics simple*, as simple as you possibly can. I can't emphasize this enough—visuals with too many things to look at, too much text, too many patterns, too many lines, are one of the most frequent of all problems in scientific talks. Remember that your audience will be looking at each visual for one minute or less (usually less), a very short amount of time. As much as a third of this time will involve tuning in to the new information shown. Therefore be considerate: give them something they can easily make sense of.
- *Design any text so that it can be deciphered clearly*, from a good distance away. This includes titles, labels, scales on graphs, tables, and so on. Bad text is also extremely common and very aggravating, showing a lack of

forethought. Good text expresses sensitivity to the situation: viewers will see that you are thinking of them.

- Similarly, *all content must be distinctly visible.* Use brighter colors for more important information, duller colors for background. Think about using the brightest colors for anything you want to highlight or make stand out. If the information is mainly visual and not textual (e.g., patterns), reduce all text to an absolute minimum. Know that people will try to read any text you put up, so if it is not entirely necessary, it will end up being distracting.

- *Try not to duplicate word graphics exactly in your own speech.* Choose the most important phrase in each listed point to recite (this can provide good emphasis), but enclose it in remarks that differ from those on the screen. Add a bit of commentary, if needed, or explanation.

- *Good slides can be used again and again,* like a resource. They can be re-used either in their original form or in modified fashion for successive talks about ongoing work. Thus, time spent on designing them well is often time saved in the long run.

Should You Read a Written Lecture?

Should you read a written lecture? There is no single answer to this question. While there are many scientists who would likely sacrifice their firstborn child rather than see this become accepted practice, many others do not feel so strongly. In some fields, true enough, it is strongly discouraged, even punished. In others, it is grudgingly allowed; in still others, however, it is fairly common and viewed with a mixture of forbearance and unconcern. In the geosciences, for example, all of these attitudes exist and are divided among various subdisciplines. Academic organizations tend to prefer nonwritten talks; those with some connection to industry are more open to presenters reading their paper.

Objectively, there is no final reason why reading your paper should be a forbidden means of communication. But—and this is a large qualifier—there are a number of trade-offs and risks involved in proceeding this way. First, reading a paper out loud can be a deadly affair for your audience, *unless* you have taken the time and intelligence to write it with many colloquial, conversational aspects. The normal scientific paper is about as far from ready material for a speech as a mathematical proof. Reading such a paper is guaranteed to bore and stun and thus offend. Second, when you read a paper, your presence for the audience is significantly reduced: your eyes and voice are directed downward. Unless you are able to lift yourself by looking up frequently and projecting well, you will tend to subside into the page (possibly for emotional reasons) and may even disappear altogether. Thus, to project an equal degree of personality, a reader must be even *more* skilled than the speaker who performs extemporaneously. Third, reading a presentation while trying to discuss visuals can be awkward, if the verbal choreography isn't well-rehearsed. All of these realities mean that to deliver an effective talk in this manner requires more work than an extemporaneous speech.

Again, there are no ultimate rules, no moral absolutes, in this area. If your field allows for such talks (as mine does), then they are a practical option, as long as you understand and adjust to the trade-offs involved. Perhaps the most fundamental argument against reading a talk is the truth that the audience has come from near and far to listen to, and interact with, the human side of science—that is, individual people, not narrating machines. A fallible, slightly anxious speaker, who may stumble now and then, but who speaks to the audience, is much more interesting than a smooth but monotonic reciting device. Thus, if you do decide to read your paper, begin and end by not doing so. Introduce your talk, and close it, by speaking directly to your listeners. Let them know that you are there, not as a hologram, but as flesh and living voice.

Lecturing to the Public: A Few Points

The topic of this chapter urges that I go on to make a few guiding suggestions about public talks. Much more will be said in chapter 19, which takes up the subject specifically. But the reader's thought and interest might be specially tuned at this point to hear a few words on this. Public speaking for scientists is once again, as it was in the 19th century, a booming industry. Public interest, however, is not entirely celebratory. Still, the result is an opportunity for researchers to become eloquent spokespeople for their (inevitably glorious) discipline, no small thing. Consider, therefore, these beginning suggestions to make your lecture effective.

Preparation

Know your audience, who it will be and how large. Keep these things in mind as you design your lecture. For larger groups, you'll need to be formal but also entertaining; for smaller audiences, more relaxed, conversational.

Know your venue, what kind of setting it is and what technology is available. Never assume there will be laser pointers, VGA adapters, water, whatever. Always find out or request. Even if the venue seems to have everything, bring your presentation in more than one form, for example, laptop and USB flash drive.

Consider: the goal of public speaking is to *connect with your audience*, not to stagger them with your depth of knowledge and imperial bearing. If you truly wish to represent your work and field in memorable fashion, you must inform and entertain your listeners, give them intellectual and emotional pleasure.

Design

Be mindful from the start that you need to *include different visuals*, more interesting ones, than for a research talk. This gives you more freedom in some ways. For example: how has your topic been covered in newspapers, on television, in the movies, in other popular places. Such coverage can provide valuable material for an opening, closing, a point of discussion, a source of humor.

Think of a central theme for your talk. Something like "where we are today in astrophysics" or "what is a genome?" works well. Try not to be too cute: puns based on movie or television program titles can trivialize a topic and make you sound condescending.

If your aim is to update an audience on your field, *consider posing your talk as an actual story*, a narration on how certain advances came about. Begin with the crucial scientific questions that remained unanswered (or unanswerable), and then show how new approaches, methods, or ways of thinking led to new findings. You can end by discussing the questions researchers now face, thus forming a neat cycle (and an image of how science sometimes works).

Delivery

Have a strong opening—something that can spark or deepen curiosity. You might choose for this purpose a remarkable fact, a striking anecdote, a quote from one or another famous source, a beautiful visual. Pick up on this opening later in the talk and, if possible, return to it at the end.

Smile occasionally as you talk. *Make eye contact* with the audience: focus now and then on a few listeners who nod back at you or otherwise indicate connection. *Move* around a bit; use your hands now and then (not wildly). Such aspects of your delivery will keep the audience attached to you and what you're saying. They like to see that you are human and alive.

Use humor carefully. Humor lubricates a message, relaxes an audience, lightens the atmosphere. One bad joke, however, will stiffen everything, grind the machinery to a halt. Think about what types of humor you might employ: try them out on others before you dare do so on a room full of strangers.

Think about using questions as transitions. These can be rhetorical: "Now, where did we go from here?" More suggestive: "How were we to interpret *this* result?" Teasing: "But if that were true, why study this species at all?" And so forth.

Add color by including any of the following elements: anecdotes or stories (make these relevant to your topic!); historical information, including images; autobiographical details (again, make them relevant); quotations from literary and/or scientific sources (use sparingly); cartoons (always check to see whether permission is needed); demonstration (use of props, a blackboard, or simply your hands); multimedia (video clips).

Anticipate and answer questions the audience may have. You might even pose some of these yourself, as part of your delivery—"Now, you may be wondering what an enzyme does." This is likely to endear you to your listeners.

Close with a strong, well-phrased ending. If possible, return to your beginning and use it to create some sort of closure. "So we see, at last, that this brings us back (full circle) to . . ." Be sure to make your closing clear and direct, so that the audience is left with something solid.

If there is a question-and-answer session at the end, make sure you repeat each question out loud, so that everyone hears (doing so shows concern for everyone). Try to keep your answers fairly short. Don't let any one question monopolize your time. If you have an insistent questioner, ask him or her to see you afterward. Cue the end with a phrase such as "we have time for one more question."

Attitude

Consider that, in some respects, *you may be donning the mantle of "the Scientist."* This means you have instant credibility, but also that you may need to convince your audience you are fully human and capable of saying things in interesting ways. In some cases, it may also make you a target of controversy (see chapter 19 for discussion of this situation).

Beware the temptation to condescend. This can come in subtle forms, or not so subtle ones. Avoid too many "Hollywood touches," such as spectacular visuals (one or two is okay; a dozen, not), amazing facts, "mysteries of nature" (this kind of speech), and the like. Also try not to pander to the lowest common denominator by falling back on clichés ("a hundred million of these could fit on the head of a pin") or simple-minded analogies—especially those of the kitchen ("in their early history, planets were like hot pancake batter") and, above all, the sports scene ("this was a slam-dunk result"). Compliment your audience by assuming they have a broader imagination.

✳

14. THE GRADUATE THESIS (DISSERTATION): WHAT IT MEANS AND HOW TO DO IT

More important work lies resting between the covers of a thesis than is often openly admitted.—CARL SAGAN

Meanings

For many scientists, the graduate thesis is the most life-changing piece of research writing they have ever done or may ever do. Such can be true both for a master's and a PhD thesis, but it is much more likely for the latter. The PhD establishes you as someone who truly has what it takes. Completing it means your student days are over; you are now a legitimate scientist. This is one reason why writing the thesis defines a challenging, arduous, multi-year effort.

The effort can be greatly helped by the theme of this book. At an early stage in your work, take the time to read through the productions of your predecessors—past theses in your field that were done at your institution, especially if they overlap with your area of research, your methods, or some other key part of your work. Consider asking your major advisor for recommendations. If examples that seem relevant to your project are few, search out dissertations at other institutions (e.g., those by students who did their research under a well-known scientist in your specific area).

These prior documents are not just for reference. They are success stories, and nothing succeeds like learning from success. They are also your friends, because they provide encouragement by showing you what works in terms of structure, organization, and style. You may find one or two that seem especially worthy; use them as companions, to listen to and to critique—finding things that you feel could have been done better (and that you will therefore know how to improve) is extremely helpful to forging your own authorial identity. You might also return to these companions for support and positive influence during times of frustration and struggle. It is more than likely they suffered dark moments, too, and managed to make the brighter shore.

Keep in mind, too, that you don't want to copy them, as templates, but adapt what they might offer, as guides. Your thesis is your own. Yes, every university is full of people who have successfully completed a PhD, and there are thousands worldwide who join the ranks every year. But your own work is unique, individual. This includes the difficulties you will inevitably face. Using the work of others is one way to make this experience less lonely. The destination is a high peak, but others are standing at the summit, waving you forward as you find your own path.

And your own path it will be. For you, the author, the dissertation becomes the experience by which you solidify and articulate your own viewpoint as a member of the discipline. This includes your stance on interpretive issues, your position on important debates, your research philosophy (hard-won, at this point), and your self-image as a "biologist" or "physicist" (you are no longer "studying biology" or a "PhD student in physics"). Writing a thesis is the one chance that many scientists have to assemble a total picture of their field, its stage of development and the directions in which it is moving, and to do this on their own—an extremely valuable thing for any scholar in any area.

Thesis Basics

In a manner of speaking, the graduate thesis is akin to a greatly expanded scientific paper. It answers the same fundamental questions:

- What did you do?
- Why did you do it? (Why is it important, and why hasn't it been done already?)
- How did you do it?
- What did you find?
- What do you think it means?
- Who helped you? (What sources did you use?)

You can think of these questions as giving your document-to-be an overall structure. But let us leave that concept for a moment.

The point is that the thesis is not an entirely alien project. True, its length and depth and years of writing effort are probably unlike anything you've done till now. But you are not on a distant, frozen world. It will be a major challenge, to be sure, and it should be. As I've suggested, it's a right of passage, the proof you can do real science. In this light, you *want* it to be challenging, because if it were anything less, you yourself would feel and probably be less genuine at the end, less prepared for a scientific career. This brings up another obvious point.

A dissertation is not merely a credential or many-page diploma. It is a result of an apprenticeship and thus a demonstration of original work, a true contribution. It is a source of new knowledge. But you are not merely generating new facts, whether from experiment or fieldwork. Your findings must be given real meaning by analysis, visual display, and, above all, critical interpretation. Such is the mark of the scientist who is no longer a student, the researcher who has emerged from the chrysalis. This comes from working independently, but with guidance. Your writing must show itself to be fully up to date in its terminology and conceptual background, yet also part of an existing research tradition. You must avoid being too derivative but also too separate from this tradition. Achieving the right balance is what all apprentices must do.

Being overly derivative is easy. All you need do is hide behind the published literature, focusing your early chapters on little more than summarizing what others have said, and populating nearly every sentence with their names. Such is a waste of your valuable time. You will find this out when your major advisor informs you, in gentle or impatient tones, that these chapters must be redone.

Conversely, being too original means, in part, leaving out too many of those names. If, for example, you find yourself writing page after page of your introduction, discussion, or conclusion without more than a small handful of references, it is probably time to stop and consider whether your committee will accept your self-deification. If this seems confusing, remember that your thesis is really about *your* work, what *you* have done and found. The published literature is needed not to take your place on the stage but to provide context, a basis for your own synthesis of the field, and to help define why your research is needed. You can get a clearer idea of the role this literature plays, perhaps, from the theses you have chosen as your guides.

Many graduate students feel that their thesis research is so specialist in nature, so confined or narrow in scope, as to interest a mere handful of people and be guaranteed as much importance as the birth of an insect. Such thinking is off the mark, not to say self-defeating. The truth is that you are making a contribution, a real one, and this is no small achievement but rather a sine qua non. In a good many cases, thesis work becomes the basis for one or more published papers that take this work further, sometimes much further, as a subject for more research and publications.

This is one reason why the "little things" must be meticulously cared for. You need to make sure that there is rigorous consistency in how you create headings and subheadings, how you label figures and tables, and how you refer to them in the text. All visuals must also be designed and sized so that their details are clear and legible. Your rigor should extend to the formatting of all text, so that the same indenting for paragraphs and spacing of lines is kept. Make sure there are no sudden changes in font style or size, for example, where you might have copied text from your notes. For your bibliography, choose a style and *stick with it*. Simply copying sources from various online articles and books will produce an annoying mess. Then there is the matter of page numbers. These, too, must be checked to ensure they go in order without any missing numbers.

All of this might seem no more than "lipstick and rouge." It isn't. Attending to these details is part of being a professional, nothing less. The reasons for this aren't mysterious. I've used the word "rigor" in the preceding paragraph not without reason. Taking care of your material shows respect toward your readers, good control over your work, and an understanding, on the negative side, that small mistakes are like missed notes in a musical performance, creating distractions and the appearance of amateurism.

Finally, a necessity: make sure you know whether your college or university has any specifications for a thesis. This might pertain to total length, format, word-processing software, even font and type size (universities can be unexpectedly particular about such things). Checking this out beforehand could save you much time and needless aggravation. Please do *not* rely on only your committee to make you aware of every such detail. Take responsibility for your dissertation from the start. After all, no one on your committee will be forced to correct such errors or suffer the evidence of an inability to follow directions.

Structure

Like any long writing, your thesis needs to be well-organized, logical in its progression, and cumulative. You will achieve this at the first level with your order of chapters. Simple enough. But a thesis, like a scientific article, is a form of storytelling: you are narrating thoughts, events (experiments, modeling, fieldwork, etc.), findings, and more. So there are second and third levels of logic and order to consider. And if you can realize these, your capabilities as a future scientific author will be much higher (you will have more sophisticated control over your material).

That being said, there are now several basic structures that a PhD dissertation may take. One of these is specific to certain fields, particularly the biological sciences, but is also used elsewhere on a less systematic basis. The most widely used structure, at present, in the largest number of disciplines, follows or approximates what I will discuss in detail below. Portions of this structure (title page, abstract, etc.) will apply to pretty much all theses, while other parts won't. I've tried to include a fairly complete range of thesis sections so that all readers will find the discussion useful.

In recent decades, it has become increasingly common for master's and PhD students to publish portions of their research, either individually or as part of a research team, prior to receiving their degree. This development is especially true in the biological and biomedical sciences and reflects a changing dynamic within these fields. It results, that is, from two factors: higher demands on students to prove research productivity at an earlier stage of their career, and a greater readiness for involving and giving due credit to apprentice researchers in significant work. Overall, this has meant more pressure at the apprentice level. It means that in these fields one must essentially *be* a researcher before gaining the essential credential that says one *is* such a species. In recognition of these realities, a nontraditional structure has become common for dissertations in related disciplines. It looks something like this:

Title Page
Abstract
Table of Contents
Introduction
Article I
Article II

Article III
General Discussion
Conclusions

The main body of the dissertation, the research portion, consists of a series of research papers that may be already published, submitted for publication, or in final preparation for submission. Each paper may have its own introduction, materials/methods, results, discussion, and reference sections, or it may not. This will depend on the institution and on the specific research involved, whether, for example, it involves a series of separate experiments or a long-term trial employing one set of methods and materials. Other aspects may vary also. If the individual papers have been submitted to different journals, for instance, their formats will be different too. As a whole, therefore, the dissertation has here evolved from a single, standardized expression of a discrete scientific effort into something more like an annotated record covering the first successful steps of a new scientific career.

Whether this situation will expand to science as a whole isn't clear, but for now the most common structure for dissertations remains close to what has held the field since the mid-20th century. In its most expanded form, this structure looks as follows:

Title Page
Abstract
Table of Contents
Introduction
Review of the Literature
Statement of the Research Problem
Methods and Materials Used
Results
Discussion
Conclusions
Appendixes
Bibliography

Notice that, like the scientific paper, the overall shape, in terms of scope, resembles an hourglass—wide and general in the beginning, more detailed in the middle, then wide again in the conclusions.

The specific structure given above is appropriate to just about any kind of experimental research. But it is neither an inflexible law nor a gospel. Many theses do not include specific sections for a statement of the research problem or a review of the literature, but instead weave these elements into the introduction. Moreover, if your research involves fieldwork or is concerned with a methodological or theoretical problem, some other adjustments might be needed.

Again, and because of this, it is *always* an excellent idea to look at previous theses done in your research area at your university, especially those under your major advisor and other committee members. This kind of "due diligence" will tell you a great deal about how flexible the structure of a thesis can be, plus many other specific things you need to know (style, chapter length, headings and subheadings, figures, tables, etc.). You might even find an example you can use as a model for your own dissertation. But don't go through other theses with this desperately in mind. It's better to remain open-minded to your own material.

Many students approach their writing in a linear way. They try to write chapter 1 first, then chapter 2, then chapter 3, and so on, to the bitter or happy end. This approach holds out a measure of comfort, like a trail of bread crumbs through a dark wood. Yet the birds may quickly descend and make off with the food. It's fine, that is, to begin with the introduction, but only in a loose way to get something down on paper (screen). Any introductory chapter or chapters will need to be greatly revised after the rest of the writing is largely done. Why? Because you don't yet know what you're going to be introducing. Thus, finalizing your intro will be one of the last things you do.

Don't feel that you can't work on your results or discussion until you've completely finished the literature review or methods and materials section. It's perfectly fine to shift around and apply yourself to different chapters in the sequence, knowing that you'll need to insert good transitions later. Writing this way can help you find connections between sections of your thesis. But if it starts to seem confusing, like having too many balls in the air, too many unfinished pieces, then go back to a more linear approach.

Writing a book-length manuscript is not algorithmic; there are no formulas. You will be developing and discovering as you go along, finding insights, making connections, arriving at new interpretations that you didn't suspect. In fact, this process—a process of discovery, just as much as any you might find in the lab—is what can sometimes elevate the task of writing to something truly exciting, an experience that will definitely help urge you onward.

One major point: *if you've already published one or more papers based on your thesis research, or completed a grant proposal to help fund it, then you have written a sizeable portion of your dissertation.* At the very least, you've produced core sections of the introduction, literature review, results, and discussion. What remains is to discuss with your major advisor how these should be expanded or otherwise modified. Your published research, meanwhile, will likely be included, either in chapter or appendix form, and will expand the prospective audience for your thesis beyond the members of your committee.

Main Parts of the Thesis

For those who have not yet published papers—the majority, in other words—the following discussion should provide solid advice on how to proceed. Keep in mind that no such advice can do the work for you, so don't expect it to move your fingers on the keyboard. If it did, you wouldn't be much of a scientist at the end. A large part of research involves putting your own skilled and adaptable intelligence to work.

Title Page and Abstract

Universities nearly always have a template for the title page that you need to follow. Your only challenge here will be thinking up a title. This is something (I believe) you should do completely on your own, without any direct how-to advice. It is the name you give this birth of yours, and it should be entirely yours. Models can help, but the first and last word here are your own.

As for the thesis abstract, this is commonly two to three pages, but can be shorter if your major advisor agrees. The basic guidelines given above in chapter 9 for this portion of a scientific paper apply here as well. You just have more space to apply them.

Table of Contents

If the title is the first thing readers pay attention to, the table of contents (TOC) is very often the second. More than the abstract, it provides a map of what the thesis actually contains. Thus, your TOC will tell the reader a good deal about how good a scientific author you are in terms of organization and some other aspects.

When done well, the TOC shows the main highways and major streets readers will be following as they progress through the text. Done less well, it either fails to deliver this much, or specifies the name of every street, alley, plaza, piazza, roundabout, and cul-de-sac. The map then becomes a blueprint; instead of providing a welcome, it overwhelms. The impression given is that of an immature author who believes that the longer and more complex the TOC, the better.

Though a fairly common problem, the overly detailed TOC can be avoided rather simply. If you find yourself creating a new section every one or two pages, such that the TOC will end up listing almost every page number in your entire thesis, you need to stop and rethink. Combine a number of these tiny (sub-)subsections under more general titles. For example, say your chapter is about the different materials proposed for the walls of a fusion reactor, and that your discussion of each candidate material requires only a page or, at most, two pages. Rather than give a separate section to each candidate, group all of them into more general divisions like metallic and nonmetallic types. If this still ends up with only a few pages under each, then place all of them in a single section labeled something like "Candidate Materials for Reactor Wall Composition."

Introduction

Finding the best way to begin your thesis can be among the most difficult challenges of all. Here you face the blankness of seemingly infinite possibility. There are, however, some fairly simple ways to commence. Some examples can help show this.

Here is the opening to a thesis on the decline of honey bee populations in the United States:

> Nearly one-third of all foods consumed in the United States consist of plants pollinated by the honey bee (*Apis mellifera* L.) ([one or more references]). The agricultural value of such pollination has been recently estimated at $18.7 billion ([reference]), making *Apis mellifera* one of the most critical species in US agronomy [no reference needed]. As a result, the decline both in the diversity and abundance of honey bees, beginning in

the mid-1980s, has generated major concerns in the food industry and led to related, yet inconclusive research among North American entomologists ([several references]).

What makes this a good introduction? It starts by providing essential context at a general level, then proceeds to make this more specific. It also establishes, beyond any doubt, the importance of the topic. And in the last part of the final sentence, it informs us of the crucial gap in knowledge that we can expect the research of this thesis will address.

Another type of introduction might begin like this:

> The aim of this study is to determine the optimal type of material that should be considered for the inner wall of a fusion reactor in future power plants. This is the first wall of material composing the reaction plasma chamber, in which extreme temperatures and levels of particle irradiation are produced. At present, however, there is no experimental facility where candidate materials can be physically tested. As a result, less direct methods must be employed.

In this case, we are given the context and the rationale for the study. Here, the gap in knowledge relates to available methods, and we can anticipate that the next paragraph will surely tell us what new methodological approach the author has chosen to pursue.

Literature Review

Whether you decide on a separate chapter for your review of the literature or to integrate it into your introduction and following chapters, it forms a central element of your thesis. The reason should be fairly obvious. You must know the relevant literature in a comprehensive way to know whether your research topic even exists or has been covered previously, whether it is meaningful and worth pursuing, and whether other researchers in the field will consider it so. But there is something more. Your familiarity with this literature shows that you have a professional level of fluency in the relevant knowledge, that you are therefore fully qualified to carry out the research your thesis is reporting on.

Keep in mind that you still need to be somewhat selective. It's a mistake to think that you should include everything you've ever read, or should have read, that might conceivably be connected to your topic. This goes for sources in other languages, too—you'll set yourself up for potential trouble by including works in German, Russian, or French if you can't read in any of these tongues. An overstuffed bibliography is *not* a sign of thoroughness. To the professional scientist, it speaks instead of a lack of confidence and the naïve and unprofessional hope that more is better.

If, however, you are working in a true frontier area where little work has been done, you may have little choice but to have a comparatively small list of sources. There is little to be gained in trying to stretch this further with irrelevant or marginally relevant sources. Instead, include early on in the literature review a statement (in humble tones) to the effect that relatively little research has thus far been performed in this area.

While there are no set rules for what to include, some guiding principles do exist. The most immediate purpose of the review is to provide essential background to your study. In some cases, it is also good to list sources that show you are thinking in original ways, but these shouldn't dominate your list. Identify early on in your research the key papers in your area, those that established the relevant territory of research and that developed it to its present level. For such authors, avoid listing those papers that simply repeat or largely replicate material in earlier publications, even if you have read all of them (a very broad rule of thumb is to include no more than around six to seven papers by the same first author). Relying too heavily on only one or two influential scientists can make you look less original, as if you're standing too much in another's shadow. Don't include sources that proved unhelpful, even if written by well-known authors, or those that were superseded by later, more important writing by the same authors.

It's a good idea to start assembling a bibliography early in the process of writing your thesis, or even before. It can be hugely helpful and can save you many hours (of algorithmic experience) to use a form of reference software, such as Zotero or Endnote, in creating your bibliography. But you need to check the results carefully; such programs are far from foolproof and can create real headaches for the unwary user. As you progress with your research, some sources may become marginal, others more central, while new papers will appear in the published literature, all of which means you'll need to continually update your list and thus your literature discussion. Stay up to date while you are working on your final document. You might even consider doing a last scan of the most relevant journals a week before turning the thesis in.

Many doctoral students put off creating a bibliography, with key sources for the literature review marked, until the main work of the study is completed. There is a *much easier* way. Begin compiling your bibliography and flagging sources for discussion even before you start work, that is, before you start your experiments, before you head off into the field, before you set your computer model in motion. Then progressively add material and reassess as you move ahead. Example: set aside a couple of hours each week, same day and time if that works for you, to update your bibliography. I can assure you that it will be quite pleasant to find in writing your thesis that you have one less authorial chore—not a minor one, by any means—to accommodate.

Statement of Research Problem and (Possibly) Hypothesis

Somewhere in the early portion of a thesis, it is necessary to discuss your research problem or question and, in some instances, the hypothesis you've adopted to guide your actual work. Some universities require that this be given its own separate section, usually following the literature review. But this is far from the rule. In some cases, such discussion comes at the end of the literature review. In others, it may form a subsection in the introduction with the title "Motivation." In still other instances, as shown by the second introduction example above, a thesis may even begin with such information.

A common way to write this kind of discussion is to break it into two parts. First, you provide a brief background summary explaining why such a study is needed, what knowledge is missing, and why it is important. You may have already presented this in your introduction, so here you simply repeat it or rephrase it in a more concise way. The second part then lays out what you did to end this unforgivable breach in human understanding. Phrases to begin this part might include "this gap in our knowledge motivates the research described in this thesis," "my research aims to solve this problem by," "the cornerstone of this thesis study is," or even "the purpose of this study is." Following this, you mention the specific topics studied and which chapters they occur in. If relevant, the location(s) of your work, where data were generated, how many experimental runs were performed, the period of time over which observations were gathered, and so on, should be included in this discussion.

If there are larger implications to your study, these too should find a place in your introductory discussion of the research topic. Later, in your conclusions, you can expand on them a good deal. Here, however, it is quite enough to just mention the larger potential importance of your work.

Materials and Methods

The fundamental reason for a materials and methods section is to show that you used good judgment and logic in choosing how to solve your research problem. Wherever in your thesis the section goes, it must be complete, detailed, and thorough enough so that anyone can repeat what you did (and, hopefully, find what you found). Thus, you should provide specifics about any equipment used, how you used it, and what materials were involved. Information on any locations used for fieldwork should be given. Similarly, details of any subjects (animal, human) involved are necessary. If you modified your methods during the study, you should note this and the reasons why.

One way to handle this section is to write it as a narrative. Tell the story of what you chose to do (your experimental design, for example) and why, how you carried it out, and whether the quality of your results validated your choice of methods. Such is a fairly simple and straightforward approach. It helps for you to keep in mind the possibility of someone repeating your work. Yes, you can leave out the part where you spilled the methylene blue on your notebook or misplaced one of the *Enteropleura* fossils you were so proud of having found. Such joys do not need to be repeated by the reader.

Some universities require you to include in this section your methods of analysis (other institutions want this in a separate section). How did you decide to organize and display your data and why? What statistical methods did you use? What levels of uncertainty are applicable? Did you use other methods that proved less informative or useful? In short, it's essential to demonstrate you know what to do with the data after generating it.

A good idea is to write a rough draft of this section soon after you carry out the actual work, while everything is still fresh. Some of the writing can even be done in stages during the data generation phase.

A final part of this section might focus on the limitations of the methods you chose. Surely they weren't perfect; no methods are, and we all understand this. But pointing out where they are strong and where they are weak or limited is an example of professionalism. Moreover, it will help head off or reduce any criticism from your committee members.

Results

The results section is where you present and describe your findings. It is *not* where you discuss what they mean; that comes, oddly enough, in the discussion section to follow. By the time you are a doctoral candidate you undoubtedly know this quite well. Yet how often does it happen that a thesis blurs the boundary between results and discussion by delving into the significance of a particular table, graph, map, visualization, or what have you, as if the author was just too excited or enthusiastic to hold back?

A results section should not be difficult to write. More work usually goes into deciding the best manner in which data might be displayed in order to make their meaning most apparent or abundant. In a fair number of cases, some part of this display will be ordained by standardized figures and forms of description (any paper or thesis on structural geology, for instance, will likely have one or more cross-section diagrams). But there is also often room for modification and originality. Such opportunities can be valuable and rewarding, but usually need some guidance from your major advisor.

Recall the earlier discussion of graphics. The demand for clarity and simplicity is particularly strong for a thesis. This is the wrong place to try to impress your readers with inventive complexity, many-layered models, data condensed (crammed) into many-column tables, or a creative sense of color combinations. Whenever you are uncertain, consult the published literature for examples.

Be sure that every table and image, of whatever kind, has a title or caption that is clear and obvious. Consistency and thoroughness in this realm speak of professionalism. Gaps and irregularities express amateurism.

Discussion

The discussion section should be the most interesting and stimulating for you to write. This is where you get to talk about what all your work has led to, what it means, why it is important, what it adds to our knowledge, what larger implications it may have for the present and future of the field (and, perhaps, beyond). This is what you have been waiting to say.

All of which is why the discussion in some ways can be the most challenging part to write. Putting your thoughts in a logical order, one that obeys the topic itself, can appear daunting when you are excited, proud of what you've found, and eager to show it. On the other hand, knowing that this section defines the final river you must cross, the last mountain you must climb, before your thesis is essentially done (all that remains is mainly summary) can also enhance such states of mind as paralysis, timidity, procrastination, even panic.

Psychotherapy or selective pharmaceuticals are not the answer. One way of approaching the task is to first restate the hypothesis you tested with your research, if this is relevant, or the primary research problem or problems to be solved. Following this, you can conceive of answering a series of questions, such as these:

- What are your major interpretations?
- How do your results support each of these interpretations?
- What are the more specific understandings or clarifications your data support? In what ways do these back your major interpretations?
- How do your findings fit with current theory in the field?
- What did you find (if anything) that was unexpected? What does this signify—why was it unexpected, and does this require a change in how the topic is understood?
- Do your data and interpretations provide important predictions, for example, regarding a particular natural process?
- What are the limitations of your interpretations? Were there areas your research was unable to elucidate?
- Are there alternative explanations, and if so, why did they seem less supportable?
- What have you added beyond previous studies?
- What are the main implications of your findings and interpretations?

Finally, a good way to end your discussion is with a subsection on suggestions for further/future work. This will obviously draw on what you have already said, not least your coverage of the limitations in your study. Point out what you consider to be the most important gap in knowledge that your work has revealed or been unable to entirely fill. A comment or two about the challenges ahead is not at all out of place.

Conclusions

At this point, you have arrived at the outer gates of completion. In this final section, that is, you are covering known ground. Mostly using new words and concise phrasing, write a summary of the principal interpretations you have made from your study—those that you would especially want your readers to remember. Do not include all of your findings; in other words, include only the ones that you feel are most important, especially significant to the science of your field.

Most concluding sections in scientific theses are quite brief, no more than two or three pages. If you also include some discussion of your work's implications and future directions for related research, then another page or two is certainly justified. You can gain help on decisions of this sort by consulting your major advisor.

That being said, it is also true that not all theses today are required to have a concluding section. This will depend on practice at your institution, of course. But if you have a choice in the matter, I would recommend you write one. Why? For two reasons. First, a conclusions section is one of the key parts of your thesis that any reader will want to consult when checking to see how important it might be to their needs. This section, after all, together with the abstract, provides the most concise overview of your core contribution.

Second, writing this section is the best way to create for yourself a kind of script to use for describing your PhD work to other researchers and colleagues. Rather than stumbling or struggling in front of others, you'll have a discourse ready in mind that will make you look like a true professional. This could come in quite handy at conferences, symposia, or any other situation where you meet new people in your field. As noted many times in this book, there is no substitute for representing your own work, thus your own scientific competence, in a confident, lucid, and fluent way. No one, after all, should be more articulate about your work than you.

But What Will Your Committee (Examiners) Look For?

So your book-length dissertation has been completed, bound, and delivered to the members of your PhD committee for their final review. What will they look for? At first, they might well lean back in their chair, gaze out the window, and imagine sunny, faraway places . . . But then they will remember that a thesis might well have some gold in it. To whit, it might teach them something new, something they need to know. It might also give them ideas for their own work, points to consider, a new technique to try. At the very least, it could bring them up to date on an area that they should probably have studied earlier but never found the time to do so. And there is that nagging question about the future: what if this student becomes prominent in the field, even celebrated? Ignoring him or her at this stage could come back to haunt both their dreams and career a decade hence.

So what do they look at first to get an idea of what the thesis is about? They read the title, then the abstract. Next, they'll glance through the table of contents, to discover how extensive a thesis it is, how many parts it has, whether they have been put in a sensible order, plus how many pages long the whole things is. After this comes a leap to the conclusions, in order to learn the main findings and whether this student can be clear and confident in his or her writing and has caught most or all typos. Last comes a survey of the bibliography to see whether he or she included the best-known sources in this area.

That's it for the first encounter. Title, abstract, contents, conclusions, bibliography. If time is short, either the contents or the bibliography might be left out, not the abstract or conclusions, however.

On the other hand, if their interest is piqued and they wish to go further at this time, they will probably turn back through the pages and glance at the figures, data tables, and any other visuals. Choosing a few of these in order, they will stop and examine more carefully a particular figure, judging its quality, completeness, and readability, then search for a couple of paragraphs that discuss it, and evaluate these similarly. Reaching the beginning again, they may read the first part of the introduction, evaluating how well the research topic is presented and discussed. At this point, if all seems in good order, the committee member will put the thesis away on a shelf, or perhaps flat on the desk, and plan to read most of the rest of it in a day or two. If all is not in order and problems have been found, he or she will commit to reading it with some care and patience.

A good thesis, then, is coherent and logical enough to have its main contents evaluated in a fairly brief time. It might have taken you years to write it, but an experienced professional can get a fairly solid idea about the quality of it in a mere hour or two. This is not an insult or sign of superficial concern for your work. The truth is that both a well-done and a poor dissertation will reveal much of their essential matter and quality in their introductory and concluding summaries, plus one or two other parts. A word to the wise can be said to reside in this fact.

One last comment: please do not think that your thesis will end up a trivial event, that the only readers it will ever have are the members of your committee, plus a (bribed) friend or relative. I have seen this claim made fairly often on blogs and other Internet postings, and it is wrong in a good many instances. No one can say who might discover an interest in your work. I myself have been shocked more than once to learn that my own thesis has been used by later students and given a reference in a number of publications. The same applies to other scientists I know, several of whom once felt the only added service their dissertations might ever find was in an outhouse. But humor and humility aside, a good many universities or university departments now routinely place their theses online and make them visible either directly or through the library system. This is especially true of public universities. Indeed, it has become true in many parts of the world, a trend that will only grow. What's more, the libraries of research universities and institutes now provide access to services like ProQuest, which collect millions of theses, on a global basis, and make them available. An essential question now is when to make your dissertation available in this way if you haven't yet published your findings.

A thesis adds to the corpus of scientific knowledge. That is its purpose and, if honestly done, its achievement. This means that no matter how minor or ancillary its subject may seem, no matter if some of its speculations be naïve, and no matter how humble it may render the trembling self-image of its creator, it retains real value. Such value exists because it is part of science itself. A high-quality thesis may not be a thing of joy or beauty, yet it is every millimeter a creation of struggle, intelligence, and contribution. It is therefore something in which its creator should, at some point, take pride. As said at the beginning of this book, adding to the great library of human knowledge can never be called a small triumph.

✳

15. THE ONLINE WORLD: SCIENCE IN A NEW CONTEXT

A spring of truth shall flow from it: like a new star it shall scatter the
darkness of ignorance, and cause a light heretofore unknown
to shine amongst men.—JOHANNES GUTENBERG

A New Medium with New Messages

A brief note at the outset: this chapter will not teach you how to use the
Internet as a scientist. It is not a survey of search engines, data archives,
e-journals, or the like. It is written for those at a graduate level and beyond,
so it assumes you are familiar with most basic aspects of Internet use, including those relevant to research in your specific field. Instead, the following sections speak to how scientists might best understand their work in
this new medium, how they might understand certain issues of importance
to online publication like intellectual property and open access, as well as
new opportunities to connect with nonscientific publics.

Let us, however, begin again with history. It shows us two things: first,
that the forms for communicating knowledge have evolved continually
from the birth of writing, in about 3300 BCE, to the present; second, that
there have also been times when specific new media—the scroll, the codex,
the printed page, and now electronic display—have appeared and changed
profoundly how people record, publish, share, and teach.

These realities suggest that the Internet will continue to develop in dynamic fashion for some time. History also suggests that the online world
will not eliminate the past, but will have to come to terms with it. We are

well past the salad days when online scholarship was going to bring a communicational Eden for all, a realm without barriers, tolls, or budget worries. This was always a fantasy. There is a great deal of the preceding age that remains part of the present—print, for example. For some time at least, the Internet will evolve alongside existing forms of exchange, changing and reorienting them no doubt, but not replacing them altogether. We can see this from another angle: in its essence, the Internet is a medium of communication, above all communication in the form of writing and speaking.

What does this mean for researchers, as people and as professionals? Put simply, every scientist should learn how to live and work online with a high level of awareness and competence. Awareness includes an understanding of what tools, opportunities, and pitfalls exist. In terms of research, this applies to every stage of work, from conception to publication. Competence means knowing, at a minimum, what uses of the Internet are expected (demanded) of scientists in their field. Knowing more than this will raise their capabilities.

Make no mistake: the Internet is far more than a communicational tool box. Like every preceding transformation in media, from paper to printing, it makes possible new forms of mind—new kinds of research, new modes of data visualization, new types of collaboration, new connections between expert and layperson. And much else. Because of this, alert and ambitious scientists will not only learn what they need to know to function well in their discipline, but also explore the online universe for further possibilities to improve and expand their research and to make it better known.

As things stand now, researchers are left to learn the Internet on their own. This is not usually a profound challenge, especially for younger scientists who have grown up immersed in the online world. Yet there is still much to learn regarding professional uses and conduct—online gaming may not be the best training ground for working in a research team with advanced data retrieval systems, for example. It would be an excellent idea for universities, particularly graduate schools, to provide courses on using the Internet for research purposes. With time, the options available to scientists will only grow and evolve further. As every field has become somewhat different in its detailed employment of online capabilities, a required class in each department might make sense.

The Positives

There can be no question that the online world has revolutionized scholarly communication, in science perhaps most of all.

Why science, above all? There are several potent reasons.[1] Every phase of research, from conception to published result, as well as archival life ever after, now depends on digital technology. Email alone now serves as the primary form of both formal and informal communication among scientists. Within certain limits (I will discuss these below), researchers can access the literature of their field, and other fields, at any time and from any location where an Internet connection exists. Science also benefits greatly from the expanded range of information it can now generate and use. The ability of digital technology to record, store, analyze, visualize, and make available immense amounts of data allows for—indeed encourages—new instruments able to generate such volumes. Satellites, large-scale accelerators, global sensor networks, and climate modeling, as well as the many faces of big data, are only a few examples. There is little doubt that we have seen only the beginning at this stage. Given its intrinsic dynamism, the Internet will give rise to new and unpredictable forms of science in the future.

Then there is the enhanced contact among researchers throughout the world, enabling new collaborations, including those at a distance. What has been termed "the invisible college" of science has therefore become larger, more international, and more dynamic than ever before, and therefore more productive. Another factor is the Internet's ability to *distribute* research in almost any form, including text, video, audio, and any type of image, fixed or animated. This is more important than we might at first imagine: the role of diverse media in science is more central than ever, and has itself become a source of new information and creativity.

Online forms of publication have also proven their power in providing new circuits for connecting with readers. A majority of journals now give space for scientists to comment on each other's work in substantive fashion. At best, this provides an open, public round of peer review, one that can lead to productive exchanges that continue beyond the pages of

---

1.   The following points draw on Christine Borgman's informative text *Scholarship in the Digital Age: Information, Infrastructure, and the Internet* (Cambridge, MA: MIT Press, 2010).

the journal and can even stimulate new collaborations. At worst, they can end in clashes or dead-end conflict between interpretive camps (even this can be useful for scientists to witness). Most often, this new dimension of contact yields worthwhile feedback that can be especially instructive to younger researchers.

Then there are the new channels that have opened between scientists and the public. This, too, carries great meaning and opportunity. Like other experts, scientists have long been trained to speak mainly, even exclusively, to their own species. In most cases, they expected their work to remain within the halls and walls of this restricted intellectual ecosystem. But online access opens this work to any number of expert and nonexpert communities, whose interest can be extremely valuable for spreading the impact of research, gaining support (vocal and financial) for future work, and advancing the cause of science more generally among the public.

Such interest will be directed to some fields more than others, of course, but the range has already come to be extremely wide. New data and ideas related to a recently discovered virus or a new factor related to climate change will gain attention from researchers in a dozen different disciplines, as well as from science writing and media outlets in a variety of nations. At the same time, new findings in such fields as ecology, fisheries, forestry, hydrology, energy studies, and climate science are being picked up by government agencies, think tanks, bloggers, NGOs, companies, and such international institutions as the UN Environmental Programme and the World Bank and incorporated into their work, including secondary and tertiary reports they publish.

True enough, the unrestrained nature of the Internet allows for the presence and rapid spread of misinformation, too. Examples of this, unfortunately, are common enough and have involved many disciplines. One of the worst cases so far involves the 2011 nuclear accident in Japan, at the Fukushima Daiichi power plant. What makes this particularly regrettable is that the sheer quantity of fabrication and hysterical nonsense has exceeded by many times the amount of legitimate scientific study made available. Scientists do have the ability—and, some feel, the responsibility—to counter through social media, blogs, and other forms of outreach. Yet such work can quickly become a full-time occupation, so *strategic* use of online networking gains much importance.

Notes of Caution

This is pretty much all good news. But, as just indicated, it also comes with caveats. The bounty of digital science comes with new barriers to access and distribution in the real world, barriers due to technological, economic, and even political factors. The Internet for science depends on several levels of infrastructure. These levels are a primary reason for why nations, universities, and research institutions around the world continue to be divided into haves, have-somes, and have-nots. Electricity is the first level—something many of the world's nations do not yet have in reliable supply. The second level includes servers (storage space and retrieval capability) and broadband. Without these elements, the greater part of contemporary scientific work and results will be out of reach. A third level comprises up-to-date computers and any peripherals that are needed for research, such as printers, plotters, scanners, cameras/microscopes (with computer connections), and so on. Next is the level of software, how up to date or dated it might be, how complete its functionality may be. Finally, we shouldn't overlook what might be termed the level of "human infrastructure," meaning the scientists themselves and their ability to utilize forms of online science, as well as crucial support personnel like network administrators, trained maintenance people, and more.

Limitations to each of these key levels of infrastructure exist in most countries today, including many where research has made major strides in recent years. The reality is that the total tends to be most complete in wealthier nations at present, therefore acting to help perpetuate the gap between advanced and developing countries. The gap is closing, though more for some nations than for others. Yet it can be widened or even created for political reasons, when sanctions affecting scientific communication are imposed on certain nations (Iran, North Korea), or when governments themselves decide to censor certain domains of research or communication of its findings (many nations). In the meantime, cutting-edge science continues to place ever greater demands on infrastructure; thus, the gap itself defines a dynamic series of phenomena. There is no single digital divide, in other words, but instead a complex hierarchy of online capability.

Researchers in advanced nations need to be aware of such realities. International collaborations are becoming the norm in a great many scientific fields, and this now regularly involves researchers from developing countries. Examples include Afghan and German hydrologists working on a study of streams in the Hindu Kush; another has Rwandan and US scientists look at carbon dioxide levels in bottom sediments of Lake Kivu. The success of such efforts, both at the data-sharing and interpersonal levels, can rely on an appreciation for differences in available research technology.

Another important caveat for online science returns us to ethical matters. The greater number of readers for online research material are other scientists, who will rarely be tempted to steal or misuse such information. But this may not always be the case. Global access to your work does mean it can be further utilized by anyone, for any purpose. It can be adopted without attribution, employed improperly for gain, even altered and misrepresented. Global access does not come with global guarantees. Malfeasance does not appear to happen very often, but it happens.

An example? Here is an ugly one, from an event I have already mentioned: A map of the Pacific Ocean showing the eastward spread of the 2011 tsunami caused by the massive Japanese earthquake was taken and relabeled as an official display of radiation coming from the damaged Fukushima Daiichi nuclear plant. The map had been produced by the National Oceanographic and Atmospheric Agency (NOAA) to exhibit maximum wave amplitudes of the tsunami, and it is a dramatic image: yellow, orange, and red fields streak across the Pacific to the Americas, with a purplish-blue "flame" (highest amplitudes) extending nearly a thousand kilometers eastward from the Japanese coast. Within the first few weeks after the actual quake, tsunami, and nuclear accident, someone stripped off the map's legend and explanation, replacing them with a radiation scale gauged to the map's colors. Months were required to reveal the fraud (not hoax), by which time it had been spread to many websites, frightening people in dozens of countries.

The Issue of Copyright

Such incidents—the exception thus far—do point up a number of issues researchers need to keep in mind. One of the most important involves intellectual property rights. Governments, including that of the United States, have decided that scientific authors should have more control over the digital content they create than they do with print. Papers, articles, images, graphics, data tables, and so forth, as well as web pages and emails, are all considered under copyright the moment of their creation as digital content ("fixed in any tangible medium," is the relevant phrase), and remain so whether held on a hard drive or flash drive. This means they are copyright protected even before they are published, thus even if they are never published. You do not need to use the international copyright symbol, ©, for this to be true (the requirement was done away with by the 1989 Berne Convention). Still, adding it does draw attention to the fact that the material is protected, so you might consider doing so if you want to emphasize this. Using the symbol does *not* require you to register the material with the copyright office.

US copyright law forbids appropriation of such content beyond fair use without some sign of permission, explicit or implied. Fair use includes quoting or other reuse for the sake of commentary, news reports, reviews, teaching, scholarly discussion, parody, and the like. Any quotation must be referenced to the original source. *Implied* permission remains a bit vague, but widely assumed. If you publish material online, you know the chances are good that if others find it of value, they will print it out or download it, use it in some fashion for their purposes, forward it on to friends or colleagues, or make some other common use of it. But if someone takes a portion of your work and represents it *as his or her own,* especially for commercial gain, he or she is breaking the law (you can sue him or her). To republish or repost it, they need your written permission. This applies just as much to things you write in email as to final peer-reviewed papers. If you wish for everyone to have free and complete reuse privileges, you can attach to the material a Creative Commons license, indicating such permission.

These comments provide a reasonable introduction to protecting your work online. Nothing, however, will wholly prevent unfair use from happening. The problem arises in that the Internet is not like anything else in the history of publishing. It is closer to society itself, in a dynamic state of expression, exchange, desire, joy, rage, celebrity, anonymity, hope for influence, and much else. Once your work is released into this seething sea, it belongs to the world; no one owns it in any total or final sense. There is no need to despair, however. All forms of publication, whether in ink or electrons, require both trust and awareness—trust that the vast majority of those interested in your work are ethical users, awareness that this might be optimistic—so knowing your legal rights is a good idea.

Open Access: Science vs. Capitalism?

We understand science to embody knowledge that belongs to humanity. Research findings aim at universal truth, or its approximation, that should be available to every interested mind. This is, in fact, a deeply rooted idea, perhaps even more deeply than we realize: who, after all, would argue that scientific knowledge *shouldn't* be open and accessible to every living person? Yet history reminds us that this has never been achieved. There have always been barriers, whether due to language, race, gender, national conflict, motives for secrecy, or differences in economic and technological level among the world's countries.

More recently, scientific knowledge has also become a commodity, an object of ownership, especially in the form of publication. Research papers, that is, are now not only career capital but also nuggets of economic value very often sold for significant profit by commercial publishers. Historically, this has meant a growing divide between the goals of researchers, which are readership and influence, and those of corporate publishers, which are sales and profit. We might at first think that readership and sales go together. But in the Internet era, researchers are not worried about how many copies or subscriptions of a journal are sold, so long as a paper can find many readers, users, and citations. Subscriptions have always mattered to science publishers, even when they were poor and proud in rags, living in academic departments. Today, however, subscriptions matter enormously, to everyone, but for different reasons.

Only a few decades ago, most journals were put out by not-for-profit organizations like scientific societies. But as costs rose for printing, distribution, and more, these publishers were unable to continue their ownership, and journals were sold to for-profit publishers. Much of this shift took place in the 1990s, but what then happened was a tremendous concentration as small- and medium-sized publishing companies were rapidly bought up by a few successful firms like Elsevier, Taylor and Francis, Kluwer, Wiley-Blackwell, Springer, and Sage. These firms, as the complaint goes, have discovered in scientific journals a golden goose: scientists require access to the primary journals in their field, so academic and professional libraries have no choice but to acquire them, allowing publishers to raise subscription rates and keep raising them even to sometimes exorbitant levels. In less than a generation, prices for prestige journals have grown by as much as 500% or more. From annual rates in the tens or low hundreds of dollars, they have headed into the stratosphere, now being in the thousands, even tens of thousands of dollars per year. The advent of online science, rather than making technical knowledge more democratically available, has aided just the opposite. A good deal of contemporary science, in other words, is now the intellectual property of aggressively for-profit corporations.[2]

Before rushing out to storm the barricades, however, we should pause to understand a few other realities. The so-called big six (companies) listed above control about 50%–60% of scientific publishing. They are powerful, but not all-powerful. A large number of smaller, usually nonprofit publishers that continue to issue important journals at much lower rates remain. Examples include the American Physical Society, the American Association for the Advancement of Science, the Royal Society of Chemistry, and Éditions Didier, plus a sizeable number of academic presses (e.g., Oxford, Cambridge, MIT, Chicago).[3] Despite the apparent oligopoly of a few behemoths, scientific publishing remains a highly diverse realm and cannot be viewed or discussed in simple generalizations, especially those that would find fault with the whole.

That being said, it is also true that the commercial giants do control a majority of high-status journals in a number of fields, such as chemistry, clinical medicine, and mathematics. As we all now know, the result has been to put the greater part of science behind a paywall. The "serials crisis" besetting research institutions throughout the world has hit academic and other professional libraries hard, with a double blow, due to escalating subscription rates *and* growth in the number of journals. In the United States and other countries, reduced library budgets, especially at public research universities, have played their part as well. "We write to communicate an untenable situation," a 2012 memo from the Faculty Advisory Council at an eminent institution begins. "Many large journal publishers have made the scholarly communication environment fiscally unsustainable." That the memo was sent to all faculty at Harvard, one of the wealthiest universities on Earth, suggests how serious the situation had become.[4] Nearly all libraries have had to be more selective in the journals to which they sub-

---

2. For actual data on this shift in scientific publishing, see V. Larivière, S. Haustein, and P. Mongeon, "The Oligopoly of Academic Publishers in the Digital Era," *PLOS One*, June 10, 2015, http://dx.doi.org/10.1371/journal.pone.0127502.

3. Most of these publishers can be found listed on the website of Publishers Global, http://www.publishersglobal.com/.

4. Full text of the letter, titled "Faculty Advisory Council Memorandum on Journal Pricing," dated April 17, 2012, can be found here: https://oaopenaccess.wordpress.com/2012

scribe, canceling many lower-use periodicals, routinely joining with other institutions to gain broader coverage, and seeking other ways to stem the tide of loss and meet their invaluable archival function. Given that this is happening throughout science in wealthy nations, how serious must the impoverishment be in developing countries?

To an ever-growing number of researchers and library professionals, the situation has an acid taste of injustice. Scientists, that is, provide the content, judge its quality, edit it, and add the status of their names and institutions, all for free. How is it that so much of the essential work is done like this, without cost, yet prices are so high and always rising? Or, in different terms, how fair (even rational) can it be that the scientific community must buy back at such rates the fruit of its own free labor? To be fair, publishers do commonly add significant value to research papers by copyediting the text, reformatting it for online use, adding links and view options for figures, performing quality control, and a fair bit more, as well as putting the entire archive of a journal online and keeping it there. All of this requires staff, time, and management, thus cost. Yet to scientists, such work can only seem like adjustment, fine-tuning. It can never compete with the level of time, effort, and thought (often matched with patience, sacrifice, struggle) that goes into the process of creating scientific content itself.

The situation has made the call for *open access* more insistent, urgent, and, to a large number of scientists, persuasive. Open access (OA) can be defined as content that is "digital, online, free of charge, and free of most copyright and licensing restrictions."[5] Its legal basis resides in the copyright holder giving consent to such access, which can be done in various ways via open-content licenses. For research, the original copyright holder is the author, unless copyright has been transferred to another entity, like a journal. OA at present tends to focus on peer-reviewed scholarship. It emphasizes that putting science behind a paywall optimizes *economic value*, but often at the expense of *research value*. Again, nothing online is free, of course. Someone has to pay for it to be put there and kept there. The question is how to do this in a way that most benefits content creators and users, those for whom the system exists.

/04/23/could-harvard-librarys-untenable-situation-regarding-journal-costs-help-move-scholars-toward-open-access/ (accessed May 31, 2016).

5.   P. Suber, *Open Access* (Cambridge, MA: MIT Press, 2012), 4. At the time of this writing (2016), Dr. Suber maintains an excellent and regularly updated resource site on open access, "Open Access Overview," at http://legacy.earlham.edu/~peters/fos/overview.htm.

OA journals, like print versions, have people on staff to do a number of necessary things: format and copyedit articles; prepare and finalize artwork; deal with journal assembly and web design; market the content; manage the review process (including, at times, the peers). Surprising though it may be, the people who do this work are professionals who expect payment for their services. As a result, just as many print journals have page charges, OA journals very often charge an article processing fee to cover their costs. These charges can vary a good deal, for reasons both legitimate and (as we will see) otherwise. A few are less than $100, but most, especially those of prestigious journals, are in the range of $1,000–$3,000. Most often, you will be charged upon acceptance or publication. Needless to say, it's always a good idea to understand what's demanded and delivered before submitting anything.

At the present time—and I must emphasize that the landscape is unsatisfactory to many and therefore evolving and temporary in its details— journals have adopted a number of different approaches for OA:

- *Gold.* The journal is accessible to readers for free from the moment of publication, *without* any restrictions or embargoes. Authors usually have to pay a processing fee. The aim is to maximize distribution and impact.
- *Green.* The journal allows the author to make available a preprint *and* postprint version of a paper but does not offer such access itself. The aim here is to give the author responsibility for making a paper freely accessible.
- *Delayed.* The journal, upon publication, is available for a specified time period (often six months) to subscribers only. After this period, all content becomes freely accessible. Such a model seeks a middle ground between traditional and OA publishing, on the assumption that revenue from the restricted time period will pay for unrestricted access thereafter.
- *Libre (Strong).* The journal is accessible for free and without most restrictions, including permissions (copyright) barriers. A goal of this approach is to allow for republication and other forms of reuse, thus distribution.
- *Gratis (Weak).* The journal makes all content free but retains copyright. This maximizes access while treating the scientific paper as a form of intellectual capital.
- *Hybrid.* The traditional journal is accessible only to subscribers, with tolls for access to individual articles by nonsubscribers, but allows an author to pay an extra fee to make his or her article available for free. The idea in

this case is that the scientific paper is a form of economic (as well as intellectual) capital.

Diversity among these models is our sign that OA matters are very much in transition, not yet in a sustainable form. We can see that from an author's perspective, the better option depends on one's personal situation. The gold standard, for now perhaps the most widespread among prestige journals, can work well for researchers whose institution will cover publication fees or who are on grants that include such fees. For those without such support, the fees are a real barrier. This includes a good many scientists, in fact, not least those at universities and colleges with smaller budgets and, again, in many developing nations. The gold approach to OA merely shifts the burden of payment and does little to control or reduce costs. As for the other options, researchers are more likely to find the green and libre approaches preferable to the so-called hybrid and delayed models. But, again, this depends on individual circumstances.

Some big moves have been made to further the grander purpose. As of 2015, more than 130 funding organizations and 580 research institutions and faculties worldwide have enacted mandates for researchers to deposit their work in freely accessible OA repositories.[6] While these numbers represent only a fraction of the total, they are large enough to be impressive and to indicate that a true OA movement is underway.

More and more universities now have policies for faculty to deposit their manuscripts in a repository run by the institution itself. When researchers wish to send the paper to a journal, they attach an addendum stating the OA policy. If a publisher refuses to consider the paper, the author can ask for a waiver of the OA policy. Such is one way academic institutions are now dealing with the problem. At the same time, when it comes to government-funded research, the OA argument has even more potency, since it involves public access to publicly funded research. Since 2013, it has been the policy of the National Institutes of Health (NIH) in the United States to require researchers who receive its funding to submit their final, peer-reviewed papers to PubMed Central, which the NIH supports. Similarly, in the UK, all peer-reviewed articles and conference papers resulting from work supported at some level by the government Research Councils must appear in OA journals.

6. Registry of Open Access Repository Mandates and Policies (ROARMAP), accessed April 14, 2016, http://roarmap.eprints.org/.

Such efforts are surely the signs of forward motion. True enough, they are not an endpoint, creating as they do a constellation of isolated digital repositories with variable contents, some holding papers, others a richer scholarly content of data, multimedia, papers, notes, and more. But they are freely available collections nonetheless, and as such embody the OA goal in vivo if also in embryo.

Will OA become the norm for all science? We cannot yet say. There are some high hopes for this, but also high walls yet to be climbed. One of the highest has to do with career capital for researchers. The push to publish in high-prestige journals, with high-impact factors, has not seen any twilight as yet and seems unlikely to go gentle into the night. More than a few established researchers continue to recommend that their PhD students build a "traditional reputation" before thinking about venturing into open access. Depending on the field and subject, this may be appropriate advice. Young scientists, unless they have cured cancer or bottled dark matter, are not in a position to challenge the existing reward system. There are other challenges to the future spread of OA, too. One of the most unfortunate, that every researcher needs to know about and understand, will be discussed next.

Yet it is hard not to feel there are greater forces at work favoring OA. As a corporate commodity, scientific publication does not sit well with its makers. Nor is this likely to change. Deep frustration with the existing paywall system has led some to take matters into their own hands. The most well-known example thus far (as of 2016) is Sci-Hub, an effort that has pirated some 50 million research papers from important journals, partly with the help of scientists willing to "donate" their personal log-in information to academic and professional libraries. Such an effort clearly reflects the widespread sense of injustice in the research community, particularly in developing nations. This is all the more true for those scientists who naively offer their private log-in info—thus exposing themselves *and* their libraries to identity theft and other mischief or damage at some future date by someone less altruistically inclined. Moreover, piracy of papers on such a large scale, though sharply aimed at the leviathan publishers, can do more injury to smaller presses and scientific societies who have kept rates low and depend on income from relatively few journals to stay alive. Stealing papers from them, which inevitably happens in a mass effort like Sci-Hub, defeats the entire purpose of the theft. Robin Hood, so to speak, dons a mask and robs the villagers.

OA surely has a future, in other words, but piracy isn't it. At present, real open access remains an idea in search of its embodiment, true enough. But perhaps some of this is inevitable, due in part to the complex challenges involved in transitioning from a world of print and limited readership to a digital universe that seems to finally offer the possibility of a universal audience.

Say what we may about the realpolitik of research competition, science bears within it profound ideals regarding the value of knowledge to humanity and its future, ideals that identify restrictions to the spread of such knowledge as oppressive. Something of this ideal was perhaps first articulated in the realm of publishing by the Abbé François Rozier (1734–1793). This Enlightenment gentleman created an influential scientific periodical (known as "Rozier's Journal") in the late 18th century,[7] arguing that science must be a "Republic of Letters," open to all. Rozier sought to rise above the barriers of language, nationality, and wealth that had come to divide science against itself, and for a few brief decades he partly succeeded, gaining contributions from many of the greatest minds throughout Europe and a readership to match. The Napoleonic Wars and 19th-century nationalism rendered Rozier's vision a noble failure in the end. Reborn as OA in a new era of global science, having survived the worst oppressions of the 20th century, the Republic of (Scientific) Letters now has potent ethical and practical imperatives behind it and seems destined to grow and find new forms that better serve the scientific community.

Caveat Scriptor: Predatory Publishers

A scholarly publication system financially based on authors paying for their work to appear runs a real risk of abuse. Even during the era of print, publishers of dubious intent happily took researchers' money to put their papers in fake or substandard journals, *especially* when such papers were of poorer quality and would not pass a diligent session of peer review. OA makes this kind of exploitation far easier to do, on a far larger scale, and with amazing impunity.

Predatory journals, as they have been called, include thousands of fake, imposter, illegitimate, and suspect scholarly publications that now pol-

---

7. *Observations sur la physique, sur l'histoire naturelle et sur les Arts* was a monthly that ran from 1771 to 1823 and that published papers by many of the most famous scientists of the day, such as Lavoisier, Berthollet, Linnaeus, Priestley, and Scheele, among others.

lute the online landscape. Many are put out by "publishers" who may be responsible for several hundred or more such periodicals and who use large mailing lists to solicit manuscripts from researchers. Most often, they have legitimate-sounding names that might, at first, fool the average researcher—*Global Journal of Pediatric Otorhinolaryngology* (a rip-off of the legitimate *International Journal of Pediatric Otorhinolaryngology*) or *International Journal of Advanced Computer Technology*. Their self-descriptions are generally quite good, in excellent English, and often quilted together from real journal websites. Do they actually publish anything? Yes, they do. Many match legitimate journals in overall regularity and size. Most claim that all submissions undergo peer review, but this is blatantly untrue in many cases, as a number of sting operations have revealed.[8] In one rather (in)famous case, an Australian computer scientist answered a repeated solicitation from the *International Journal of Advanced Computer Technology* by sending a paper consisting entirely of seven choice words repeated many times: "Get me off Your Fucking Mailing List." Several weeks later, an email arrived stating that the paper needed some newer references but was otherwise excellently suited to the journal and was accepted, pending payment of $150 (no discount for irony). The episode, however, did not end happily: the paper was withdrawn, and the mailing list remained unchanged.[9]

There are no universal signs by which these journals betray themselves. On the contrary, it is also the case that some long-lived and legitimate periodicals have been sold and then changed into predatory versions by the new owners. Where are these scams located? A good many have been traced to such august headquarters as a suburban house outside of Reno, Nevada, or a small warehouse in northern Mumbai. A detailed look at several dozen of them indicates a broad range, from complete fraud to legitimate but lesser-quality science. Some of the latter even belong to pub-

---

8. See, e.g., E. Segran, "Why a Fake Article Titled 'Cuckoo for Cocoa Puffs?' Was Accepted by 17 Medical Journals," *FastCompany*, January 27, 2015, http://www.fastcompany .com/3041493/body-week/why-a-fake-article-cuckoo-for-cocoa-puffs-was-accepted-by -17-medical-journals. The most extensive such sting operation, involving hundreds of journals, is detailed in J. Bohannon, "Who's Afraid of Peer Review?" *Science* 342, no. 6154 (2013): 60–65.

9. "Journal Accepts Bogus Paper Requesting Removal from Mailing List," *The Guardian*, November 25, 2014, http://www.theguardian.com/australia-news/2014/nov/25/journal -accepts-paper-requesting-removal-from-mailing-list.

lishers owned by the big-league for-profits, like Elsevier and Kluwer. Thus, watchdog lists of such journals describe their predatory nature as "potential, possible, or probable."[10]

The only element shared by all such journals is poor-quality science, including pseudoscience. In a few cases, this may reflect unreliable peer review (peers who don't do their reviews) at a legitimate journal, but far more often it seems to come from the motive to make money, scam the system, and therefore not bother very much or at all with quality control. A large number of authors of these papers are in developing countries; though they, too, must publish to advance, for a variety of reasons their training, equipment, funding, and so on, may not (yet) be on a level commensurate with authoring papers of the quality demanded by most international peer-reviewed journals. Do the "predators" therefore fill a gap, even satisfy a global need? This is a major question, not easy to answer. From the viewpoint of the individual researcher, the answer is almost certainly yes. But when we consider the long-term career needs of these scientists, plus what is required for research to thrive in their countries, we must say no—publishing inferior work helps no one and, in fact, works against real progress.

And the matter does not end there. Predatory publishers invite academics to present their work at conferences, almost always for a large fee (your first hint), and to serve on editorial boards. The former brings in cash, the latter legitimacy. Such conferences do take place, and their programs usually list important people and many talks. When you show up, however, you find none of these scientists attending, and instead of hundreds only a few dozen presentations, with weak coffee and stale pastries on hand. Some conferences are more swanky than this, but the quality of scholarly material is noticeably low and disappointing and may even be sketchy. You soon realize this is an imitation; you are in academic zombieland.

At the last, predatory practices are not a natural effect of OA. Making this link is like blaming higher education for the 19th-century "entrepreneurs" who schemed to get rich by buying a cheap house on cheap land and

10.   At this writing, the best-known and possibly most complete list of predatory journals is kept by Jeffrey Beall, a librarian at the University of Colorado. His catalogue of publishers and stand-alone journals can be found at "Beall's List: Potential, Possible, or Probable Predatory Scholarly Open-Access Publishers," last updated April 14, 2016, accessed April 14, 2016, http://scholarlyoa.com/publishers/.

opening a college with high tuition. Scholarly predation today is mostly a scam directly tied to the realities of authorship, those that require researchers to publish and to pay for such publication. Researchers everywhere tend to agree that OA defines a much-needed alternative to toll-based science, an alternative that should become more dominant in the future. We must hope that its evolution finds a business model able to better integrate quality control.

Social Media: The New Culture of Science

One of the most unexpected yet immensely fertile developments in the online world has been the growth of social media. For scientists, these media constitute tools of tremendous importance. Indeed, they are probably more important than many users realize. Why do I say this? For a simple but weighty reason: social media provide science a much larger, visible, and accessible presence in the world, and also a highly strategic one. To nonscientific audiences, social media have the ability to make science more abundant, immediate, and, in many ways, *friendly*. I shouldn't have to explain the advantages that can flow from this (but I will anyway).

A main benefit of the Internet for researchers is to increase their individual presence. This means contact with both scientific and nonscientific audiences, but it also involves the ability to *create* such audiences. A good many scientists, of course, have found it valuable to set up their own websites, either about themselves or their research specifically. It is common for universities and other research institutions to provide scientists some amount of online space, whether this be a page or two or an entire website. Depending on what this allotment may be, you might wish to set up your own separate site as well. If you are a fairly confident writer, or wish to become more of one, creating a blog might be worth considering.

Beyond these options, the possibilities open up still further. With Facebook, Twitter, Tumblr, and more, researchers are able to multiply their visibility many times over, as well as their networking, sharing of information and interests, seeking of answers to all manner of technical questions, and so on. These tools—and undoubtedly others still to come—are the essence of the new online culture of science. They are dynamic, evolving, unpredictable in their varied futures. But taken separately or together, they all embody something undeniable: the vibrant, energetic desire of scientists to connect with one another and with the greater world.

Blogging

Blogging can be a successful career move. It can also be an engaging hobby or a trip to vanity fair, but my concern here is with more serious goals.

Using a blog to talk about your field and your work, its value and possible importance, is reasonable self-promotion. In another era, we might have looked upon this differently, as shameless braggadocio. Those days are melting into memory. To be sure, scientists trained in pre-Internet times may still feel this way, at least initially. But many have seen the good that blogs can do for science, linking it to public interest and support in an era when science denial seems all but contagious.

Some researchers worry that blogging about one's own work can lead to bias, glory seeking, and misrepresentation of a discipline. Undoubtedly, this happens; science is too competitive for it not to. Yet the tendency can be easily avoided by a bit of ethical behavior: telling readers when you're discussing your own research, when you're offering an opinion or interpretation (not a final fact), even when you might be wrong. Not doing so while trashing the work of others isn't just a matter of airing dirty laundry; it's indecent exposure. Blogging means acting as a representative of the scientific community. You may be right in what you say, but you will be far more convincing and successful in promoting yourself if you say it with dignity and respect.

Beyond such considerations, it makes little sense to put a gag order on any form of articulate enthusiasm about your research—what you have chosen as your life's task. We all agree that researchers need to network, cultivate their options, make their work known and appreciated. To borrow a barb from Thomas Edison, "We can't be like the German professor who, as long as he can get his black bread and beer, is content to spend his whole life [locked away] studying the fuzz on a bee!"

The blogosphere on science is galactic in scale. There are thousands of researchers and science writers, as individuals and in groups, as well as popular science magazines, government agencies and labs, science museums, and so on, who have their own blogs. None of this should dissuade you in the least from starting your own if you are drawn to the idea, whether as a mission or experiment. Warnings aside, an effective blog can have huge benefits. Greater awareness of your research gives it more impact, with potential positive effects like invitations to collaborate, new publication opportunities, public lectures (paid and unpaid), media attention, all possibly leading to a gleam in the eye of grant committees and academic departments with job openings.

Of course, none of this is inevitable. Some of it (e.g., media attention) may be unwanted. Exposing your ideas to the outside world also brings a degree of risk; not all of the commentary you receive, especially if you work in a controversial area (climate change, stem cell research), will be laudatory, or even civil. Giving opinions, weighing in on controversial topics, making connections between your field and its implications for larger social needs or problems, all of this can open the door to both angels and demons. Perhaps the greatest risk of all, however, is that blogging can become an infinite time sink, even an addiction. Staying in control is key. The challenge of saying no to demands arising from praise and appreciation can be met only by the ability to keep priorities and responsibilities in clear view.

An essential fact: blogs have become the primary global form of science writing (*not* scientific writing), including science journalism. Indeed, one of the most productive ways for researchers to use a blog is to hone exactly this communicational skill of explaining real, complex science to nonspecialists. Skill of this kind is no mean achievement; in real-world terms, it is a form of added power, an enlarged ability to inform and influence the minds of others. Having it also means you do not have to rely on any interpreters or intermediaries to deliver your ideas, findings, or data.

To set up a high-quality website and blog is not an overwhelming task, nor is it a lazy afternoon's errand. There is much material online to provide you with advice, both detailed and concrete. But before diving into any of this, search out examples that you find particularly good (yes, models)— both in appearance and content—and take note of why you like them. Perhaps you already follow some. It may take a little time to identify what you like most, but it's well worthwhile. Then talk to other researchers in your department or institution who have blogs and mine their experience. It will take a little time to put everything together, but the effort is more than worthwhile.

In the end, though available to anyone, blogging is not for everyone. Being an articulate and informed spokesperson for your field and your own research can be a great thing, but it involves trying to put yourself on stage, in public. That is not everyone's cup of tea, especially in the sciences. At the same time, to blog or not to blog is a decision that can be undone or reformed. You can always, at any time, call a halt, cut back, redo a blog. Or decide to start one.

Social Networking

Blogs will remain an option for researchers, but other social media, particularly those focused on social networking, are likely to become a necessity. Why is this so? Simply put, social networking has become a primary means of connection among human beings. Nothing in the history of communication has expanded so rapidly, globally, and pervasively. More *homo sapiens* now use social media to communicate with groups than any other method, be it voice, email, telepathy, or what have you. Such communication conquers distance, time, and context, though not (yet) language. Roughly half the world now uses one or more social media sites, with the number of unique users growing yearly. Like so much in the online world, social networking has proved to be a highly dynamic, evolving realm.

How does all of this relate to science, specifically? In the first place, scientists have been rather reluctant to take social media as a serious domain of benefit, even though it is clear that these media already saturate non-face communication in both advanced and developing nations. Such reluctance appears to be rooted in several ideas—for example, using such media isn't scientific work, rather it is only for self-promoters, and it takes up too much time. Such ideas are false and easily dispelled (as we will see). Some of the problem may be due to misperception. Informal talk among senior researchers indicates a tendency to view these media as merely online chatter valves or as a form of popular culture best employed by celebrities, politicians, and other mirror-loving types. While such uses definitely exist in abundance, they no more define the possibilities of social media than firecrackers and perfume delimit the capabilities of chemistry.

Since 2010, social media (e.g., Twitter) have revealed a capacity to be at the heart of key world events. They helped spawn and spread revolutions in the Middle East, mass protests in Russia, Ukraine, and the United States, and social and legal changes of a high order in dozens of nations from Brazil to Korea. As a result, they have also become the subject of government censorship, spying, propaganda, and other efforts at control. We could spend much time and space talking about such developments, but their ultimate meaning is irrefutable and plain: social media are forms of power. By "power," moreover, I do not mean force or authority, but instead direct communicational reach, contact, and influence. Such is why educated professionals in many disciplines have adopted these media to create new informational markets, new channels of persuasion, and much more.

But let us return to scientists. Like others who have used social media on a serious level, whether in academia, government, or another sector, you can use these media to

- Keep informed about the latest research, science-related news, global developments, teaching ideas and innovations
- Create new relationships with those in the same or related fields
- Exchange ideas, data, papers, images, and so forth, cleanly and immediately, without the limits experienced with email
- Foster new collaborations
- Combat misinformation and misunderstandings of technical facts, events, and issues
- Create interdisciplinary groups for discussing specific issues, problems, ideas
- Generate new research questions and plans for international projects
- Follow conferences in real time (with limits)
- Follow any number of key journals
- Share and discuss such information with interested colleagues
- Send out job announcements to targeted groups

Yet beyond all of this, another dimension of use exists, one that can give you the powers of strategic communication. Using social media to make contact with nonscientists, that is, allows you to

- Generate attention for your field
- Generate attention for you and your work
- Create interest groups focused on your field and your own research
- Expand your presence by acting as an informed commentator on important developments, ethical issues, controversial topics, topics in the news related to your expertise
- Act as a distributor of accurate scientific knowledge in general and on specific topics in which you are knowledgeable
- Reveal the human side of science (e.g., express real excitement at a new discovery, outrage at a bad decision, naked enthusiasm for a proposed project, grief at the loss of a famous colleague)

Some readers may feel uncomfortable with the apparent self-aggrandizement of these goals. But step back a moment and reflect: what are the benefits of having more of the public—a domestic and international public—interested and engaged in science, in your field, in your research? Most of us would readily agree that science, given the forces of opposition to it, needs all the spokespeople it can find. Aside from the added value social media can bring to your scientific work, they hold out the offer of acting as a frontline advocate for your life's work—no small thing, by any means.

At this writing, the two most globally used social networks are Facebook (>1.5 billion users) and Twitter (~330 million users). Twitter is widely employed around the world, except in China, where it has been banned for political reasons (China has several social media sites of its own, Baidu, Weibo, and YY, while Russia has Vkontakte). Scientists have generally found Twitter simpler, more direct, and useful for actual science-related connections, reserving Facebook for more personal (family and friends) material. Beyond these two platforms are several others that offer different functionality: for example, ResearchGate (requesting and sharing of scholarly papers, conducting brief discussions, finding collaborators); Instagram (photo and video sharing); Tumblr (multimedia posting with short-form blogging); LinkedIn (professional networking, business-oriented). Undoubtedly, the total range of these media will continue to evolve and expand. Future options could well be geared to specific groups, like scientists or perhaps even individual disciplines like physics or the life sciences.[11]

Because it has proven particularly useful for scientists, Twitter deserves some further discussion. It is mainly a real-time networking, information, and microblogging platform. Messages (tweets) are a maximum of 140 characters (not words), but can include embed links, images, and videos. Twitter lacks a file-sharing capability, but there are free online services for this. There are also many "Twitter 101" websites to help guide you in the early stages. The basic process involves signing up (free), creating a profile (be honest), and then searching for topics, people, and subject areas you want to engage. Initially, it helps to observe how other scien-

11. At the present time, ResearchGate (www.researchgate.net/) approaches some aspects of a science-specific platform, yet it is used by academics in the social sciences and humanities too. It has a tiny fraction of the users that Twitter now commands mainly because it is largely confined to scholars. This situation may be highly suitable for researchers who do not wish to engage a larger public. But it is therefore limited since it lacks the advantages of connections outside the scholarly sphere.

tists use Twitter, what they post, who they follow, and so forth. At some point, you'll want to begin replying to messages, receiving replies, and starting your own conversations. A definite goal is to explore content using hashtags. These are content markers with the # (pound or hash) sign as a prefix to a key word or phrase associated with a specific subject area, like #chemicalengineering or #urbangeology (no spaces allowed). Clicking on a hashtag or using it as a search term will deliver all tweets containing that tag, thus presenting you with much content and many other people to engage with. Beginners can first explore hashtags in the messages of other scientists, then those in their own field and research area. The greatest challenge is dealing with the volume; even the hashtag for a particular conference can bring you a flood of information and messages. Understanding this, and the discipline needed to remain selective and in control, gives you a real leg up.

Seems fairly simple, and it is. But there are important matters to keep in mind. One of these is who you want to be online, that is, your persona, in terms of the language you use. Most people write their tweets as if they were talking informally, sentences flying out of their mouth. For scientists, however, it often pays to be more restrained: gaining an online reputation as someone who is informed, insightful, and helpful counts as a worthy aim. Similarly, provocative statements are one thing, but researchers who wish to retain credibility should avoid inflammatory, accusatory, or otherwise offensive statements. Like all social media, Twitter can be subject to abuse by "trolls," who try to disrupt or derail conversations, especially on newsworthy or debated topics. Far less of this happens for scientists than others, but it still pays to be aware of the possibility and to keep your cool and dignity if it happens to you.

One final point, a request actually: when posting on Twitter or other social media in the role of scientist, please do your best to sound intelligent and thoughtful. Keep in mind that your messages are not spoken; no matter how short or breezy, they do not evaporate but are inscribed in the electronic version of stone. You can delete them from your own profile page, but not from anyone else's if your message has been retweeted, a common occurrence. Impressions you create remain, like a dent in metal.

Thus—advice I repeat in other chapters—don't write silly or embarrassing things or use metaphors that make you sound ridiculous. A recent example, about a surprising discovery on Pluto: "It's like going to Neptune and finding a McDonald's." We can't all be great wits, of course, able to unsheathe a glinting phrase whenever the occasion demands (keeping a list of such phrases might not be a bad idea, though). But we can at least be attentive to our role as representatives of the scientific enterprise, especially in a public setting where our words may easily be taken and repeated elsewhere. Or, as once put by the eminent Samuel Johnson, "Every man, in whatever station, has, or endeavors to have, his own followers, admirers, and imitators, and has therefore the influence of his example to watch with care."

✳

*Part 3. Special Topics
in Communicating Science*

16. FOR RESEARCHERS WITH ENGLISH AS A FOREIGN LANGUAGE

L'auteur dans son oeuvre doit être comme Dieu dans l'univers,
présent partout et visible nulle part. (The author must be
in his text like God in the universe: present everywhere
but visible nowhere.)—GUSTAVE FLAUBERT

English as the Language of Science: A Few Realities

"Science and technology now have a true global language, and it is English." How true is this statement? Very true. It is a truth that comes with both responsibilities and limits.

To be fully active members of their profession in a global sense, scientists must learn to read, speak, and write English. Knowledge of this language is not only necessary; it provides a set of potential opportunities that did not exist until very recently. What kinds of opportunities? The chance to collaborate with other researchers worldwide, in more than a hundred nations. The chance to receive advanced training in the most scientifically sophisticated of these nations. The chance to apply for jobs in any of these places, not only in academia, but in private industry, NGOs, intergovernmental organizations, and more. None of this is guaranteed, of course. In fact, the only guarantee is that without English, almost none of it would be possible.

As the global language of science, English is helping globalize science. For centuries, modern scientific effort was largely confined to a handful of wealthy countries; that period of history is now over. Contemporary science belongs to no one group, culture, gender, or nation. This is only one reason why a global language is a good thing for science. There are others that have been discussed, weighed against the disadvantages, and found convincing.[1]

The necessity to learn English may seem a burden, especially to those from less wealthy nations where science must struggle to keep pace in many other ways, too. But history reveals, beyond any doubt, that this kind of situation has been a core part of scientific development since very early times. People drawn to science had to learn Greek or Chinese or Sanskrit in ancient times in order to have access to the most sophisticated thinking. In the early medieval period, it was Arabic (or Chinese again); then from late medieval times to well after the Renaissance, Latin. During the 18th century, Latin gave way to French as the tongue of advanced international learning. From the 1850s to the First World War, knowledge of German provided access to the most advanced methods and theories, especially in the physical sciences and medicine.

Scientific work today is mainly expressed in English for international audiences. Yes, there are other world languages, too, in which science is published: Spanish, French, and Russian, for example. But none of these are dominant, as English is, in Europe, Africa, Asia, Oceania, or North America. The spread of English has only grown with the Internet, including the use of social media. Not only are the great majority of influential journals in English, but international conferences, symposia, and other meetings are now routinely held in this language.

That being said, what English are we talking about? American? British? Indian? East African? These all have significant differences, as any linguist will tell you. To be "safe," however, scientists should focus on learning to read, speak, and write what is called British or American English, as this will help guarantee the widest access to the written and spoken material of science.

Read, Listen, Speak, and Write

Please note the order in which I have listed these skills—read, listen, speak, and write. Reading is generally the easiest ability to learn. Listening (and understanding what you are listening to) is more challenging, but there are many ways to practice this, from television and movies to online videos

---

1. S. L. Montgomery, *Does Science Need a Global Language? English and the Future of Research* (Chicago: University of Chicago Press, 2013).

from conferences in your field (a good idea to use). Speaking, at a working level, is difficult because you are now actively using the language. It is most difficult if you are afraid to make mistakes and therefore rather shy. But because English is a global language, *most* speakers make mistakes; you should never let this stop you from talking, since every chance to do so is an opportunity to practice.

Writing, for scientific publication especially, is hardest of all. Why? Not because it is more complicated, but because it involves three kinds of skill: knowing English fairly well, knowing the specific English of your field, *and* knowing how to write at a good, functional level.

The standards for a scientific paper are high, but they do not demand perfection. To write a publishable paper, you must follow the format used by the journal and be able to do the following:

- Make your document clear and readable (no confusing parts)
- Use terminology correctly (the right terms, where they belong)
- Make each sentence understandable to a native English speaker

"Readable" does not necessarily mean totally free of errors. While you should always try to eliminate all of these if you can, editors are more forgiving than they used to be. This is a recent change in policy for many journals, and it is a good one; some good science was rejected in the past simply on account of nonstandard (non-British/American) English. But— and this is very important—you cannot use this as a guarantee. The level of editorial leniency depends entirely on the journal. A paper that is halfway between English and another language (Singlish, Japanglish, Spanglish, and so on) will not be accepted by any legitimate periodical.

Reviewers and editors generally won't reject an article for some minor mistakes if the science is strong. However, errors that make your paper difficult to comprehend are likely to kill any chance of its publication. Even one bad paragraph or a dozen poor, unclear sentences can do this. Remember that, unlike speech, writing cannot be changed once it is in print: it is there forever and for all to see, again and again (and again). To write for publication is therefore more demanding, as a skill, than reading or speaking in professional situations.

Learning to Read in Order to Write

How does one learn a foreign language? First, by imitation, and later, by improvisation. We learn to copy sounds, to memorize words and rules, to repeat phrases and patterns until we can create our own combinations. We study examples of good expression, how their sentences go together, and we try to follow their example. In short, we learn to reproduce the language in its native forms until these forms become native to us (or at least partly so). It works the same way in writing science.

Learning to write begins with reading. Once you have a reasonable knowledge of English—you can read this chapter, for example—you need to learn the vocabulary of your field. This means the terms and technical phrases that are commonly in the most recent textbooks and papers. At the same time, you see *how* this vocabulary is used. You see it in articles and reports by scientists whose first language is English or who write English well. You absorb the "dialect" of your field so that it becomes part of you, and you can express your own work in it later on.

How do most scientists acquire this "dialect"? Over a period of years, constant reading of the literature helps develop a sense of what sounds right and what doesn't. The process works for most scientists, but it takes time. It is very often slow and sometimes incomplete.

It can be done, however, much more quickly and efficiently. How? By a type of focused reading that uses selections from the scientific literature. The process is simple; it costs next to nothing and can be carried out anywhere. But like all things related to language learning, it requires discipline.

This process is based on finding and using examples—examples of good-quality writing in your field. You will see from the following the basic idea behind it. I have organized my suggestions into a kind of program that is easy to follow. You may, however, adapt it in any way that works well for you and your circumstances. Here is how to begin:

Make a file of excellent articles in your field that you've come across in your study or reading of the literature (e.g., those you most wish you could write). You can do this on a computer or print them out, but having paper copies will make things easier.
- Make sure these articles are not too difficult for you to understand.
- Do not copy articles because they seem complex or because they are

written by the best-known researchers, but because they are clear, well-organized, and understandable.

- Ask your colleagues/professors to recommend any papers they think are especially well-written and well-organized.
- Make sure these articles are recent, preferably less than five years old.
- You can also use chapters from recent books, reports, or other publications.
- If possible, have a colleague who is a native English speaker look over your selections and make any recommendations.
- Keep the total number of selections fairly small, between 10 and 20.
- Above all, try to find one or more articles on subjects that are as close as possible to your own work.

Reread these examples of good writing on a regular schedule.
- Take 30 minutes or so of each morning or evening to do this.
- Concentrate on one article per week (or longer).
- Take time, now and then, to listen to the words as you read them silently.
- If your memory is strong, try memorizing chosen paragraphs.
- Keep a vocabulary list of terms you needed to look up, and review this every time you reread the paper; for each entry, write out the sentence in which the word occurred.
- Eventually choose one or two articles whose topic is close to your own work and that might act as a guide to help you write a paper of your own.

Imitate your examples of good writing.
- After your daily reading, take a few minutes to copy out paragraphs from your chosen article.
- Copy the same paragraphs for several days at least.
- Do this for an article whose subject is very similar to your work.
- Try to add a few sentences of your own (using fake information) to each paragraph or replace a sentence or two within the paragraph.
- Try to write a new paragraph in the same style as this article, using your own data (or fake data).
- Have someone else, preferably a native speaker, read over these short writings and comment on them.
- Think of these activities as part of your language training.

These activities all have a single goal: to help you develop a sense of what sounds right in the language of your field and what does not. You may not have time to do all the things listed above. At the very least, however, it will be an enormous help to read and study good examples of scientific prose—examples that you have chosen yourself, or that have been recommended by colleagues, and that you find simple, clear, and worthy of imitation.

Helpful and Unhelpful Worries

It is a very common belief among scientists who are foreign speakers of English that the more they know about grammar, the better they will be able to write. In other words, the more rules they memorize, the more mistakes they will avoid and the better their usage of English will be. My students come from all over the world, and many, many of them tell me this is how they are taught to learn English all through middle school and high school.

But the idea is incorrect. Learning to write well doesn't happen this way. Knowing the rules of music notation will not make you a pianist or composer. Knowing all the rules of soccer will not make you an excellent player. The ability to write well comes from a different type of understanding, a kind of trained ear for good expression. This is true for writing in your native language; it is even more true if you are working in a second or third tongue.

Many books try to teach good scientific writing through rules and examples of proper usage. Most of these books are for native speakers of English; a few are not, but contain the same material in slightly simpler form. Nearly all of them claim that the literature of science is badly written and desperately in need of emergency care. They imply, therefore, that it is a mistake to imitate the writing of other scientists. Such is an author's way of saying that you should rely on his or her set of rules for good usage. In other words, it is an advertisement. But it is something else as well. Stating that scientific writing is bad, in general, is a way of telling you not to rely on your colleagues in this area. It also suggests that nearly all editors are poor at their jobs. Neither of these ideas is true—nor could they be, unless science were about to disappear from the face of the Earth.

It can be a help to read through and study a book or two on scientific style and usage. Such books can help train your ear in a certain way, guiding you away from some common errors (I especially recommend Vernon Booth's *Communicating in Science*, second edition, because it is short, clear, and practical).

But this kind of book will never actually *teach* you how to write. This can only come from close, intimate contact with what others have successfully produced, especially in your own field. Good prose, not a rule book, will be your best teacher. You will find, moreover, that it is much easier to learn acceptable writing from the published literature than from a guide on scientific style.

The Value of Colleagues

Science is very social in its work. Scientists interact with colleagues at almost every level and on a constant basis. Some of this interaction is formal, as in seminars, meetings, conferences, and so on. But most exchanges are daily and informal: when we have discussions in the lab or field, drop in at someone's office, chat in the hallway or on the phone, go out for lunch or a drink, exchange email. A great deal of science gets done in these ways.

Colleagues are very often our most important resource. This is especially true when it comes to writing. You *must* have others read over your work if you are going to submit it to a journal or other publisher. These individuals, whether friends or fellow workers, will be your first editors, and you should think of them this way. If at all possible, get a native English speaker to look over what you've done and let you know how it reads, where it needs improving. Ask your readers to evaluate your grammar, style (readability), and organization (how well ordered are your sections and major points?).

Remember, however, not to ask too much of anyone willing to read your paper. For example, don't request that an English-speaking colleague correct every grammar or spelling error. Request that your readers spend only as much time on this task as they can safely spare—for if you are too demanding, they will not want to help you in the future. It might save time to show them several early sections of your paper first, rather than an entire draft. Their comments will help you improve your writing throughout the later parts of the paper, thus minimizing any changes you'll need to make there.

There is one final source of help you can use, if you are at a university or research institute. These institutions usually have a writing center, whose staff is there to assist others, including undergraduates, graduate students, and faculty, to improve their writing. The students who work for a writing center are usually very friendly and eager to help, and they do so for free.

A Bit of Practical Advice

Most of the earlier portions of this book contain advice on how to put to-gether a scientific paper, report, or proposal. The chapters are simply writ-ten and should help foreign speakers of English, too. Here, however, are a few essential points that might be emphasized for the sake of such re-searchers:

- If you are writing a paper for publication, you first need to choose a jour-nal or journals where you wish to submit your work. Study several issues of your first choice closely. Pay attention to the organization of articles, their length, and their style. Copy the instructions to authors and refer to these as you write. Obey all style guidelines.
- Be conscious of whether articles in your journal (and field) must follow a given order of sections (e.g., introduction, materials and methods, re-sults, discussion, references). Many fields, and therefore journals, do not have this type of standard.
- If you find it difficult to start writing, find an article on a very similar sub-ject and rewrite the opening paragraph using your own specific topic and information. Remember that this is a way to help you begin; it is *not* a method for writing an entire article (using it this way may bring accusa-tions of plagiarism). If you do copy an entire paragraph or two and re-place only a few words, go back later on and rewrite it so that it is more your own.
- Anytime you are uncertain of how to say something, or when you are stuck and can't seem to write anything more, take out one of your article examples (on a subject close to your own) and read it over several times. See if there are any phrases or sentences you can use to help get you past your sticking point.

The Necessity of Patience

Learning scientific English is part of learning English generally and re-quires patience. It may take as long as a year following the type of program discussed above before you just begin to feel comfortable writing the lan-guage of your field. You will be working to get better and better for years afterward—just the way native speakers do. Writing good scientific papers is no easy thing. Indeed, good scientific writing in English is quite foreign to many native speakers, too—that is the reason for a book like this one.

Please note that it has been shown, time and time again, that learning a language is best done when study is performed often, a little at a time, on a regular basis. Massive doses of studying and memorizing once a week (or even less often) are ineffective. So is studying for hours and hours every day. This is likely to be even more true with regard to writing, which demands a particular ear for language that comes from long-term, thoughtful exposure.

Improving your writing will require patience and time. Drowning yourself in books and articles will not make the process happen more quickly. On the contrary, it may lead to disappointment when you see that results do not come as rapidly as you expected. Repeated study of single good examples is the best way to train your sense of what sounds correct in the writing of your field. Think of it as a form of training, where repetition is crucial.

17. TRANSLATING SCIENTIFIC MATERIAL: GUIDING PRINCIPLES AND REALITIES

Writers create national literatures . . . but world literature
is written by translators.—JOSÉ SARAMAGO

Perspective

Translation has been called the second-oldest profession on the streets of authorship. In fact, it was there at the beginning, creating marriages among texts and peoples. Indeed, it helped give birth to new fields of knowledge and learning in all the world's cultures. The idea that this act of transference brings about an inevitable degree of infidelity or loss is more in the way of an ancient literary prejudice than an eternal truth. It has never had much relevance to science. Here, translation has long been an invaluable source of dissemination and progress.

Mobilizing knowledge about the natural world from one language to another, thus from one civilization to another, defines an activity of immense historical importance. Such is no less true for mathematics. If planetary astronomy came to Europe via the Greek, Syriac, Hindu, Persian, and Arabic tongues, so did it arrive in English, French, German, Italian, Spanish, and so on, via Latin. If this seems too vague or venerable, consider the matter of how Newton's works and Darwin's *Origin of Species* achieved their worldwide influence. The very first issue of *Nature* (November 4, 1869) carried a number of translations, including that of Goethe's writings on nature by Darwin's great defender, Thomas H. Huxley. There is no avoiding the simple fact that modern science rests on the pillars of translated knowledge.

Today, science has a global language—English. What does this mean for translation? Leaning on intuition, we might think such work has become vestigial, without need or necessity. Yet the very opposite is true. A global lingua franca greatly expands the demand for translation. This is because the greater part of research work and publication continues to occur in national languages, so it must be rendered into English if it is to be shared internationally. There is also the flip side. Given that the most prominent international journals have increasingly become English-only, their material must be rendered into local languages for the sake of those not (yet) proficient in this global tongue. Technical translation from English into Chinese, Spanish, Arabic, Russian, and so forth, has therefore grown tremendously. Beyond academe, moreover, governments and companies are vast producers, consumers, and dispensers of scientific information. All of this activity, though silent, even invisible, to the main conduct of research, involves the daily transfer of rivers and mountains of material.

The importance of translation will remain strong, therefore, because of the globalization of science. And this is something we should welcome: every nation deserves its own scientists, engineers, and mathematicians, just as it does its own writers, artists, and musicians. Translation makes this more fully possible.

Should we then consider science translation a part of science itself? Absolutely. Such is just as true today as in eras past; indeed, it may be even more so (see the next section). Translation is also, therefore, a very worthy profession and a necessary one (this author has himself been among its yeoman and written of its importance[1]). Invisible it may too often be, yet were it to stop overnight, many wheels would soon grind to a halt.

The Field Today

Science translation should be understood in terms of the demand for its services and the state of the profession to satisfy this. Experience shows the greatest demand lies in the private sector. In the era of globalization,

---

1. S. L. Montgomery. *Science in Translation: Movements of Knowledge through Cultures and Time* (University of Chicago Press, 2000).

international firms must conduct business in multiple languages; the need to localize (as it's called) material around the world is imperative. This is no less true for companies that deal in scientific information and hardware. Firms that manufacture, sell, and buy products worldwide related to bio-medicine, computers, electronics, IT (information technology), military and defense, agriculture, construction, and other technical areas have great need for translators. International business-to-business communication today often involves scientific and engineering subject matter, whether concerning a piece of medical equipment or the patent on a chemical process for shampoo. Energy and mining companies, whose work is global and who need to have documents rendered into local languages, form another sector of demand. Since the 1990s, medicine has become one of the largest domains for science translation around the world. But there is also growing need for translation in certain niche areas, such as waste management and pollution control. Such topics will change and so should be investigated.

Government ministries and organizations are another source of demand. The European Union spends on the order of $8–$10 billion annually on translation and interpretation, because of the requirement that important documents be made available in the languages of all member states. Some of these documents are highly technical in nature and so require science translators. This need for translated scientific material has grown for many government entities worldwide. An obvious example, relevant to nations as different as Afghanistan, Mongolia, and Bolivia, has come from new proposals for large-scale mineral extraction and the corresponding necessity of government bodies to create environmental regulations and impact assessments.

Research scientists have their own set of translation needs. Besides the transfer of whole documents from one language to another, there is frequent need for translation of sections of texts, whether part of an article, report, or book, as well as figures, diagrams, maps, tables, and other visual materials. Some of this is for collaborative work, but some relates to teaching and lecturing. Scientists whose English is not strong may seek translation of their own writing in their native language so they can submit the result to an international journal. This can also involve a script for a talk, lecture, or presentation to be given at an international meeting in English. Until all scientists are fully competent in English—or whatever language may come to replace it as the global lingua franca—these kinds of translation demands will have a daily place in the doings of science. Such demands also extend to email, web pages, social media, and other Internet material. Here the need relates not only to English as the target language but many national tongues.

Those languages most in demand for science translation, other than English, are international ones—Spanish, French, German, Arabic, Chinese. Unsurprisingly, the most growth in recent decades has been concentrated in East Asian languages. Japanese remains very important (both English to Japanese and vice versa), with demand for Chinese and Korean increasing rapidly. Other languages of note include Turkish, Russian, and Portuguese. At the same time, a very successful career can be built on translating into or out of less international or so-called niche languages such as Polish, Kazakh, Urdu, or Khmer. Then, too, there are the languages of nations undergoing rapid economic development, where scientific effort is on the rise and thus the need for translation as well. At this writing, such nations include Indonesia, Vietnam, Thailand, Mongolia, Uzbekistan, Mozambique, and Ethiopia, among others.

How are all these demands for translation being met? First and foremost, by a large, expanding, and fairly well-paid, worldwide labor force of science translators. Some of these people are employed in companies and government organizations, but most inhabit international translation agencies and work as professional freelancers. Science translation composes a subset of what is now called the translation services industry, estimated at around $35 billion in 2015, and growing.

Agencies have expanded in size, number, and specialization quite rapidly since the early 21st century. They now compete in a global market that has the potential to grow a good deal more—for example, into Africa, Southeast Asia, Central Asia, and the Middle East. They have also expanded in the number of languages they work with, thus increasing their client base and the market as a whole. It is now routine for professional translators, in agencies or freelance, to become expert in one or two particular areas, such as medicine or military and defense. In the 1980s and 1990s (when I did translations), this was much less the case, on account of a relative paucity of translators with scientific training. Specialization testifies to the huge market growth, the resultant increase in such translators, and the requirements for good-quality work and quicker turnaround. Once a realm of scattered artisans, science translation has increasingly become a domain of journeyman professionals.

Indeed, many graduate programs now offer degrees in translation, especially in Europe and Canada, and, to a lesser degree, in the United States, Australia, and Russia. Some of these programs have direct contacts with government bodies and translation agencies that allow them to help place graduates in job positions. Be sure, however, that any such programs you investigate are not strictly devoted to literary translation. Also determine whether they include any training in, or exposure to, the use of translation software or other computer-based aids. These elements are now a normal part of the professional translator's tools.

Some Points about Computer Translation

Most of us have likely tried one or another kind of translation software, like that offered online by Google. And, to put it nicely, we have likely found the results to be variable. Useful in some cases, the end product can be indecipherable in others and sometimes badly misleading (I have seen German versions of my own work that twisted certain points into quite creative shapes). Overall, it seems fair to say that such tools will not replace human translators anytime soon.

It's true that the field of machine translation, or MT, has progressed greatly in recent years and is now able to do some remarkable things, particularly with spoken language. Some companies working in this area, like Microsoft, are now able to render simple English into other languages in the same voice. But such capabilities only work for limited types of discourse, above all ordinary speech and writing with a modest vocabulary. For something as complex as scientific texts, where much of the meaning relates to knowledge external to the text, held by the reader, MT has mostly proven to be unequal to the task. While it may accurately render some sentences, particularly those that aren't too rich in jargon, it can't be trusted to produce accurate or, in some cases, even readable translations.

The upshot is that for professional translators, a more specialized type of translation software is needed, particularly one that can be fine-tuned for individual fields. Even then, MT versions will require careful proofing, editing, and, in some places, retranslating. We can envision the day, perhaps, when human translators act more as interlingual editors. But that day is far off at best.

Advice to Beginning or Possible Future Translators

Translation defines a vital and indispensable part of science communication. As a translator, or translator-to-be, you may not *do* science in the research sense, but you are a mover and distributor of science around the world and are thus more central to the globalization of scientific work and knowledge than often recognized.

It is essential to understand the nature of the task. At base, translation is *not* about pouring a text from one flask into another. The task of the translator is to create a *new* text in the target language. Experienced translators know a literal rendering of any original is impossible. No two languages have a one-to-one correspondence in vocabulary, grammar, or syntax. If they did, machines would do the work quite nicely (early MT trials in the United States, focused on scientific material in Russian, found to their Cold War dismay it couldn't be done). Yes, technical terminology has word-for-word equivalents in most of the world's major languages—"T-cell receptor" has its counterpart in French, Russian, Chinese, Japanese, and Arabic—as we might assume (and hope). But beyond this, all translation involves interpretation; decisions have to be made about phrasing, sentence structure, punctuation, and a great deal more. In the actual doing of such work, the translator is never a mere servant of the text; she or he is also its author.

The goal of every good science translator, therefore, is to make the result look and sound as if it were originally written in the target language. This is the very best you can achieve; it is an ideal of sorts, but a practical one that can actually be attained. Such achievement will always depend on your level of experience, your own writing ability, and (no surprise here) how much time you have to work on a project.

Beginning translators can benefit from a few other points. First—this may sound like I'm breaking down the open doors of common sense, but there is a good reason for it—if you wish to do this kind of work, make sure your ability in the relevant language or languages is very strong *and* you have learned a significant part of the technical vocabulary used by the field in which you want to translate. It requires care, and often experience, to understand that while complementary terms may exist in different languages, as noted above, these can take many forms that include loan words (usually from English, but not always), calques, literal translations using existing words, metaphoric translations, new coinages, and all of these.

New translators should not be confused or distracted by all this variety and complexity. Creative rendering of a German term like *Einsprengling* as "sprinkle," because it's derived from the verb *einsprengen* (to sprinkle), is not only wrong but revealing—it shows that the translator is either not qualified or too lazy to look up the word in a scientific dictionary, where "phenocryst" would be found. Good digital dictionaries are an essential tool for every translator, of course, and are even available for free online.

It is all too common for people who are not ready (they may have had scientific training and know a foreign language moderately well) to try to launch a translating career. The consequences are often unfortunate, even improvident and damaging, for all concerned because of wasted time, money, and reputation. Such is one reason why agencies employ only experienced translators or give incoming candidates a qualifying exam. Second, recognize that you need to be a competent (preferably good) writer to be a capable translator. This, too, should be obvious, but isn't. Poor writers make bad translators. Bad translators do not last very long; nor should they.

What about approach? Because it is science, you must understand every technical point and detail in the source text. Accuracy is critical; even small errors can degrade a translation significantly (it has been said that no errors are ever minor in technical translation). Therefore, first read quickly through the source text to get an idea of how easy or difficult it will be for you to work on. Be aware that your evaluation should include terminology, scientific content, and writing style of the source. You need to decide whether you can complete the work in a reasonable amount of time, or whether it will overly tax your capabilities. Clients or agencies will usually allow you to evaluate a document first, so you have a chance to say no. If, for whatever reason, it seems too difficult or demanding, don't accept it; otherwise you'll risk making errors and damaging your credibility for other work. If it looks doable, make special note (mentally or physically) of any parts that may require extra work. Go back and read the first few paragraphs more slowly to determine how good or bad the writing may be—a poorly written original can end up requiring a lot of added effort.

When translating, it's a good idea to always read a paragraph or two ahead. At the very least, you should read each paragraph in its entirety, carefully, before working on it. You need to grasp the order of its parts, the logic that holds them together. Also note any transitions used to link paragraphs or major points in the text. At each step, when you read a paragraph or series of paragraphs, try to listen for how they might sound in the target language, if written for a similar kind of document. This may require you to delete some small parts, like repetitions or other redundancies, and perhaps add others. Don't be afraid to make such changes if they add naturalness and clarity.

This brings us to a fundamental question. Should a translation be strictly loyal to both content *and* style? In other words, should it accurately render the scientific substance and the writing, no matter how poor? Or should you consider rewriting any parts that are badly expressed, confused, repetitious, or otherwise distracting from the content? Arguments have been made over the centuries for both sides. Every translator must deal with the question, consciously or otherwise. Doing so while conscious is better.

Here, then, are a few points of professionalism to consider. If you deliver a translation with poorly written passages, muddled meaning, or other problems, your work may well be rejected and the original sent to someone else who better understands the nature of the client relationship. Fervid explanations claiming *fidelitatem et honorem* (fidelity and honor) will not advance your cause, or career. Translating science has its profound aspects, without doubt, but it is also most often a commercial activity. As such, it demands a *useful* product. Bad writing, whatever its cause, is an acid that corrodes utility. You only have to put yourself in the place of the client to understand why. Meanwhile, the best translation agencies often, or even routinely, have technical translations put under the eye of a reviewer, who will not allow defective prose to go through (and may append a note that is not in your favor for further work).

As for final rules about how much change to perform on a poor or mediocre original, there are none. It is a judgment call, therefore a window on your translating ability. Skilled professionals make decisions on a flexible basis, varying the degree of change as seems appropriate to each sentence, paragraph, section, or document. For beginning translators, or those new to science translation, this will be challenging at first. The goal, again, is to make the final result accurate in every detail while sounding native in the target language. No higher definition of a useful technical translation exists.

Can models be of use? As for most things dealing with authorship, the answer is yes. Having on hand a paper or report in the target language whose topic is the same or very close to that of your project document can be really useful. It can clarify points of usage, appropriate phrasing, use of terms, and more. Sometimes its greatest aid can be to get the proper flow of language moving in your head, so you have a better sense of how your translated result should sound—a type of benefit that shouldn't be under-estimated. Thus, building a file of such reference works, whether scientific papers or website pages, is a good idea, especially for beginning translators.

Final Words of Encouragement

Translating scientific material is noble work, despite what may appear to be its mercenary and utilitarian aspects. By noble I mean that it takes place within an overall historical context of great antiquity and venerable impor-tance. Such importance, moreover, has never been higher than it is today. Nor has it ever been relevant and beneficial to so many parts of the globe, to so much of humankind. These realities are not often articulated, unfortu-nately. But they are no less true for that.

On the remunerative side, science translation, part of the larger cat-egory of technical translation, is often the most highly paid of any trans-lation work. This is not due to its nobility but rather the skills required, which are especially high in language(s), subject matter, and the ability to write a clear sentence. While demand for these skills has grown, only a small number of professionals continue to have them all. Science transla-tion then shouldn't be taken up lightly, wishfully, or indifferently. It is seri-ous work and comes with serious rewards—the ability to live and work in different parts of the world, and to move between a freelance and corpo-rate existence.

The range and sophistication of aids to translation are destined to ad-vance. In some ways, professionals can look forward to their work becom-ing somewhat easier to perform, more efficient in use of time. Yet such aids, be they related to machine translation or to new types of dictionaries, will never eliminate the core need of human judgment and interpretation. No less than in literature, the task of the translator in science will never be that of a servant or retainer.

＊

18. MEET THE PRESS: HOW TO BE AN EFFECTIVE AND RESPONSIBLE SOURCE FOR THE MEDIA

As a vessel is known by its sound, whether it be cracked or not;
so men are proved by their speeches whether they be
wise or foolish.—DEMOSTHENES

Why This Chapter Is Different

For more than 25 years, part of my regular work involved occasionally acting both as a provider of technical information to media outlets and as a writer of articles and other publications for such outlets. I was both a source and a reporter, in other words. Some interesting tales could well be told about this bimodal experience, but I mention it here as a basis for a single, essential truth: for a great majority of scientists, dealing with the mass media is one of the most challenging, promising, confusing, enraging, and potentially rewarding encounters of an entire career. It can also end up being trivial, even after everything just mentioned.

For these reasons, I have structured this chapter rather differently from the others. Part of my intent is to introduce the reader to media realities, the media's demands and expectations, and how they both overlap (in some part) and conflict with those of scientists. To extend this information into a solid, usable awareness, I have provided much advice in the form of recommendations—how to prepare for an interview, how to be both sensitive and savvy in the encounter, how to view your role as spokesperson, how to evaluate the result. Far more advice is given this way than can be demonstrated by examples in a single chapter. The advantage, I believe, is that, in a fairly short space, such advice can help readers form a sensibility about the science-media interaction. This interaction, after all, can only be a charged and challenging one in any democracy, seeking as it does a communicational embrace between powerful, expert knowledge and the popular mind.

Facts and Issues

Some scientists I know love to tell war stories about dealing with journalists. At conferences and parties, they trade tales as if showing scars in a steam room. When pressed a bit, however, these same researchers will also admit that coverage in major newspapers, in news magazines, and on television will likely increase citations and even funding. Pressed a bit harder, they will also confess to having witnessed informative journalistic pieces, including magazine articles or documentaries, that have treated a familiar topic with both pith and panache. Mixed attitudes toward the press, in fact, are common within the scientific community. And vice versa. The encounter, however, is too often wrongly portrayed, remembered, or feared, as a high-injury contact sport. It is frequently the very opposite, particularly if you are prepared for it (which this chapter will help you become). And in any case, it is necessary.

Scientific knowledge has enormous power in the contemporary world and a large part of it is paid for by the public. Yet, unless translated into more ordinary language, this knowledge is all but inaccessible to the vast majority of people (including scientists, if we think of a botanist trying to comprehend geophysics). Most scientific work, meanwhile, has goals that overlap with public benefit—to advance knowledge, increase material wealth, support national and institutional prestige, underwrite professional success, among others. Moreover, science is involved in burning realities of immediate social importance: medical research, energy, environmental protection, and so on.

Given these basic truths—power, arcaneness, public dependence, and social effect—it's only to be expected that science would eventually become the topic of sustained and warranted coverage in the press throughout much of the industrialized world. Such, indeed, is the present state of affairs: science is news.

But not always frontline or front-page news. Since the late 1990s, the mainstream press has greatly reduced the number of journalists and other staff devoted to scientific topics. This has partly been offset by the periodical literature devoted exclusively to news about science and written by professional science journalists. Weeklies and dailies, such as *Discover Magazine*, *ScienceDaily*, *Physics World*, *LiveScience*, among many others, have maintained a fairly stable readership.

Yet the largest growth in science writing for the public has come from the online world. Here, the forms for explaining new research results, government science policy, controversies, and more are now many and diverse. Researchers, academic departments, NGOs, government labs and agencies, international programs, and professional science writers have all created websites and blogs and are using social media (e.g., Twitter, Facebook) to try to inform, as well as sway, public interest.

More information about what is happening in, and to, science is available today than at any time in history. A significant part of all this new coverage is good-to-excellent in quality, though it can sometimes put a mask on opinion, even bias, and call it fact. The reason for such complexity is not only intellectual tug-of-war among researchers ("for every PhD, there is an equal and opposite PhD"). Other motives, chiefly the desire to attract positive attention, may be involved.

Yet, despite online options, a majority of the public continues to get its news about what is going on in science from major newspapers, television reports, or science-based periodicals—that is, science journalism. Such is one reason researchers remain divided on how to think about press coverage. Many do remain suspicious or uncertain, and feel encouraged in this by the coverage given to deniers of climate change, dubious science like cold fusion, and bad science like the false link between vaccination and autism. Scientists in general live on details; they understandably tend to view a sound bite about research as an oxymoron. More than a few have turned away from dealing with the press altogether to use social media. This is a mistake.

The realities of the present situation therefore demand a certain intelligent awareness on the part of scientists. Social media will not replace science journalism anytime soon. To simply dismiss this journalism as popular pabulum, or shun it as invasive and debasing, would be both arrogant and self-defeating. We should understand that this journalism is not only for the nonscientific public; it is for scientists too, who read it to keep informed about other fields, controversies, and more. At the same time, viewing the press solely as a platform for promotion—for selling one's own work, department, or field—is also misguided, and an excellent recipe for trouble.

Realities of Journalism

The media define a vast, complex series of institutions. The people who make up these institutions are as mixed in their competence, savvy, and interest as are those of any other, equally sizable "estate"—like science. Even more, journalists work under a range of constraints that *must* be acknowledged and understood, for they determine in part what appears in their reports. To deal with the media effectively and intelligently, therefore, it helps to be aware of certain realities involved in the encounter.

Here is how a successful and well-known science journalist begins a book on her profession:

> I work as a science journalist. That is, I pay attention to interesting and important developments in science and engineering, talk to the researchers who uncover them, learn about the ideas behind them, and then communicate this information to the public as engagingly as I can.
>
> I like this work not just because research is fascinating, though of course it is, but also because scientists and engineers are interesting. Typically, they are passionate about their work—and passion is an attractive quality.[1]

This passage comes from Cornelia Dean, a senior science writer at the *New York Times* who served as the paper's science editor from 1997 to 2003. It tells us a good deal. It does not tell us what, perhaps, some of us might want to hear: that journalists are shallow, fame-hungry creatures out for a story on world-shaking breakthroughs, controversies that can be hyped to sell, or a sensational exposé on scientific fraud or failure.

But no, the quotation above informs us of something very different: that the media has a focus on *people*, the human side of science. The story that journalists wish to tell is very different from the one researchers create in their papers. Emotion, personal struggle, persistence, controversy and heat, challenges of a new idea or theory, such are among the things that might be used to nucleate a tale. Knowledge is important, but the individuals who create it and debate it and, in some cases, inflate it are just as interesting to the public, sometimes more so.

1. Cornelia Dean, *Am I Making Myself Clear? A Scientist's Guide to Talking to the Public* (Cambridge, MA: Harvard University Press, 2009), 1.

What emerges from this fact? That scientists need to be prepared to talk about the human side of their work, and that, if they want their science to be handled well and thus handed to the public as such, they need to help reporters, not shun or talk down to them.

The requirement for such a perspective becomes even clearer if we look at the demands a science journalist must face. They are far from simple. Rendering what is essentially an elite, expert knowledge into terms accessible to Mr. T. C. Mits (The Celebrated Man in the Street) involves much more than merely giving the emperor a new wardrobe. It demands something closer to a *rewriting* of scientific knowledge, adapting it to a wholly different context of presentation and audience. This is no simple achievement, if done well. Moreover, one must add to this challenge the realities of contemporary journalism, which include (but are not limited to) (1) writing on deadline, (2) limitations of space, (3) editorial control, (4) pressures for simplification, (5) demand for definitive statements (from "experts"), and (6) the need to attract interest.

How do reporters cope with all of this? Through a host of choices regarding type of material, level of detail, use of metaphor, and various narrative techniques. While editors have a lot of say in the choices their reporters make, established conventions also exist. Here, we might want a more detailed answer to the question, what makes science news? Another well-seasoned practitioner of the art has responded by providing us a list of criteria, in order of weight: fascination value, size of possible audience, importance, reliability of the results, and timeliness.[2] The significance of this list will not escape most readers here.

Who are science journalists? They range from full-time reporters on science to those who cover "the science beat" only occasionally. Some have scientific training, possibly at a graduate school level. Others, however, have migrated into science journalism from political, economic, or feature reporting. There are, too, the various types of publications and media for which journalists work. These aim at different audiences. For example, some periodicals are targeted at professional scientists; such would include *Science* and *Nature*, both of which have news sections. Others target both

---

2.   B. Rensberger, "Covering Science for Newspapers," in *A Field Guide for Science Writers*, ed. D. Blum and M. Knudson (New York: Oxford University Press, 1998), 7–16. For those scientists who might (or should) be interested, it is well worth the time needed to read through this official guide, endorsed by the National Association of Science Writers, to get an idea of how "the other side" lives and works, how they conceive of their craft and its subjects, including scientists.

professional and amateur science watchers, for example, *Science News*, *Scientific American*, and the *New Scientist*. These very often—though by no means categorically—present favorable coverage of new work, but have increasingly taken to discussing controversy, fraud, public consequences, and political contexts as well. On the other hand, newspapers like the *New York Times*, the *London Times*, *Le Monde*, *Frankfurter Allgemeine Zeitung*, and *Asahi Shinbun* are likely to show mixed styles of coverage, though in recent years the overall balance has been far more pro than con in tone. The broadcast media, television and radio, are most interested in capsule reports that contain some nugget on a new discovery, public health issue, controversy, or the like (health is the preferred topic in the United States). This branch of the media is, by far, the most constrained in terms of space and time and so looks to hook the viewer/listener often more aggressively than do other news outlets.

What about narrative techniques? These have tended to vary over the decades, too, but return to several standard journalistic approaches (science journalism, be assured, is a subset of journalism generally). One of these involves beginning with a "grab", for example, "If Martin Grossbauer is right, we may all be living on insects by 2050," and then going on to explain and amplify the startling/enigmatic opening. Another technique is to start with a pun or play on words—"Astronomers like to think dark thoughts, particularly when it comes to questions of matter and energy in the universe." There is, too, the traditional "four Ws and the H" method (who, what, where, when, and how), which resembles the standard news story. Since the 1990s, meanwhile, it has become fashionable to adopt techniques from feature writing, in which articles put a personal crisis or transforming experience at the center of the story: "Mary Johnsen had never heard of interleukin 2 before she was diagnosed with deadly hepatitis B."

Such are but a few examples. Obviously, much more could be said about these techniques, and the images of science and scientists they offer. The important thing for researchers is to be aware that the journalist is a craftsperson, who, in the midst of various pressures and constraints, constructs a story in particular ways, just as any writer must.

In the end, two great forces pull and tear at the science journalist—the same great forces that rend all of journalism, but that act here with greater intensity. These are the demands to *engage* (win interest, entertain, fascinate) and to *inform* (offer knowledge, increase understanding, urge awareness). These demands take place within an institutional setting where economic realities are often pressing, even paramount. Our capitalistic system of making and selling news means that any such outlet must attract attention—lots of it—to succeed and survive. This is not a negotiable reality.

Ideally, of course, reporters are third-party observers who work consciously in the public interest. In practical, day-to-day terms, however, they must satisfy their editors, whose job it is to sell newspapers, broadcast value, whatever (and thus keep themselves and their reporters employed). The realities of contemporary journalism ensure that science correspondents, however essential to an informed democracy, only occasionally view their main responsibility as educating the public. This is perhaps unfortunate, but largely inevitable given the way things work. A crucial point for scientists is to avoid the lure to blame reporters for these realities.

It must be said, however, that in the eyes of the scientific community, the news media did themselves a poor service with their handling of climate change denial in the early 21st century. For more than half a decade, the American media in particular continued to give credence to a small number of voices stating that the science on this topic was "uncertain," as if the research community lacked a true consensus and continued to hotly debate the matter. Such was patently untrue; the data indicating warming trends, sea-level rise, melting glaciers, ocean acidification, changing ecosystems, and the high probability of human cause were settled matters. Attempts by the media to present a "balanced" view by quoting "both sides" revealed an inability to differentiate between scientific research results and claims aggressively made by vested economic and political interests. The idea of "balance" in this case seriously misrepresented the truth. True, the situation was made more difficult by certain climate researchers making charged public statements about how socioeconomic reality had to change (in a way that favored environmentalist sensibilities). But many journalists failed to do their homework and simply reported on a scientific controversy claimed by the deniers but which didn't exist.

The problem, thankfully, was corrected, though not without much pressure and lobbying from the global scientific community. In the end, it brought to the fore a fact about media coverage: it is a basic principle for reporters that they not become tools or mouthpieces for a particular viewpoint, so they tend to be wary, even suspicious, when a scientist claims there is only one possible interpretation for data with major social implications. This is something researchers might keep in mind.

A Brief Comparison

Here is another paragraph by a science journalist describing his trade:

> When I tell a tale, I like to write it as though it were a film . . . creating
> close-ups, medium shots, and long shots through words, so that readers
> can visualize the story and feel that they are there, watching. But you can
> only write those revealing close-ups and backgrounding long shots if you
> did the reporting that brought you the material to work with. Good writ-
> ing, with its need for fine detail, eloquent quotes, and vivid imagery, de-
> pends on good reporting. There are no shortcuts. Be willing to spend the
> time it takes to get to know the facts of the story, the characters in the
> story, and the issues the story raises.[3]

Again, we see that "the facts" make up only one thread in the larger tap-
estry to be woven. But here we have more detail than in the earlier quote.
There are "characters" and "issues," "eloquent quotes" and "vivid imagery."
Human interest and social significance, not knowledge alone, form the
pilot subjects of the journalistic effort.

That being said, let us now look at some rather sobering words by a
noted sociologist (Dorothy Nelkin), who has herself written eloquently on
how the press covers science:

> Journalists convey certain beliefs about the nature of science and tech-
> nology, investing them with social meaning and shaping public concep-
> tions. . . . Was interferon a "magic bullet" or a "research tool"? Was Three
> Mile Island an "accident" or an "incident"? . . . Is dioxin a "doomsday
> chemical" or a "potential risk"? . . . Are incidents of scientific fraud "in-
> evitable" or "aberrant"? Some words imply disorder or chaos; others cer-
> tainty and scientific precision. Selective use of adjectives can trivialize an
> event or render it important; marginalize some groups, empower others;
> define an issue as a problem or reduce it to a routine.[4]

---

3. M. Knudson, "Telling a Good Tale," in Blum and Knudson, *Field Guide for Science Writers,* 77.

4. D. Nelkin, *Selling Science: How the Press Covers Science and Technology* (New York:
W. H. Freeman, 1995), 11. This work should be required reading for every scientist. Also very
good on science and the media are the essays collected in E. Scanlon, E. Whitelegg, and
S. Yates, eds., *Communicating Science: Contexts and Channels, Reader 2* (London: Routledge,
in association with the Open University, 1999).

The press does not merely provide stories. It also acts as political broker and agenda maker. Thus, for scientists, it helps greatly to read press accounts with a critical eye, noting how material is ordered, what language and imagery are used to present it, what levels of sophistication are assumed (or not assumed). Science writers have their own considerable power to shape public sensibility and establish menus for public policy. But they are not alone in this process, by any means. Consider their sources.

The Motives of Scientists

As a species, scientists are no longer categorically shy or passive in the face of media attention. An increasing number have transformed themselves into spokespeople, public debaters, even regular commentators. There is also the demand, brought to bear by many sources—scientific societies and research institutions among them—that investigators should actively reach out to the public, make their work comprehensible, available. Whether this means education, promotion, or popularization, however, often depends on local circumstances. The scientist of today does not so much descend from the gates of Olympus, as emerge into the spotlight from less elevated precincts.

It does not require a public policy expert to suggest that positive exposure can lead to many benefits—increased influence, enhanced funding, job opportunities, an advance into favor for one's own discipline. Conversely, negative publicity can be very damaging to a program, career, or area of research. Scientists today realize that the stakes involved in media coverage can be very high. Science journals, too, are very aware of the advantages media publicity can bring. Some (e.g., *Science*, the *New England Journal of Medicine*) provide advance copies of their upcoming issues to journalists, who then cover articles in them, making such journals appear the key sources of new science, which they therefore become. This has certainly worked to mutual benefit for researchers and reporters. But it can easily backfire too. As already noted, the press has its own needs and agendas, and experienced journalists have their antennae up for attempts to exploit them.

The complexity of the situation and the stakes involved have led research institutions, including universities, to create media or public information offices, complete with one or more public information or science press officers and associated staff. Such officers commonly have experience in print, radio, or television journalism. They are rightly considered a valuable resource for any interaction with the press. They can act as an advisor, intermediary, filter, trap runner, or mediator for scientists—indeed, it is a good idea to seek their input any time you feel unsure or need information about dealing with the media. They may well come to you, if you've published something important, served on a high-level committee, or done something that makes your institution look good. Many institutions' public information offices publish an online daily or weekly celebrating the achievements of faculty.

Public information offices (PIOs) are not new, but have grown to become a normal part of the interaction between the research community and the public. It is now routine in many cities and university towns for local or even national media to place a request with a PIO for an "expert voice" on a topic or issue that needs to be covered. PIOs often make lists of such "voices" to use in these cases. If you have any interest in acting as one of these, contact your PIO and tell them what subjects you can speak and write about (yes, being such an expert can mean writing an op-ed or article, too). In historical terms, the PIO represents an attempt by institutions to take positive control over how their knowledge producers are portrayed, or portray themselves, in the common culture. They have become an added player in the longstanding struggle for the public's interest. Journalists, to be sure, understand this very well. They much appreciate dealing with people who understand their craft and business. But the relationship is professional, not cozy. The PIO is not there to hook you up with one of its buddies at CNN.

Let us admit that, as scientists, we too are torn by competing forces. These, to be sure, are not the same as those for the media. We want favorable coverage for the labor of our hearts and minds, a clean and shining public image. Yet our training and the ethics of our profession tend to make us suspicious and even defensive toward public exposure. Much of scientific work is aimed at priority demands, being at the forefront of a particular area, and scrutinizing others' work with a critical eye. The mentality that goes with this is one of caution, keeping things quiet and proprietary until published, distrusting large claims offered without supporting detail. Moreover, scientists are trained to view themselves as separate from the nonscientific public in particular ways. We usually view our own knowledge as exciting, valuable, even admirable—otherwise, what's the point?—whereas to the press, it can qualify as mere "subject matter." The realization of this truth can be disillusioning.

Scientific publication and media publication are therefore two very different things. Researchers should never measure the second by the standards of the first. No journalist would ever hope to find the converse.

The Public's Interest

The public is interested in scientific work for a great variety of reasons. Science is powerful, expensive, elitist, inaccessible, yet also forward looking, optimistic, full of promise, even, at times, spectacular. More than any other area of knowledge, science carries with it the sense of advancement, moving ahead, exploration, newness. Science visibly improves its own powers, adds to itself, and carries us all forward, in general feeling, with it. There have long been moral and emotional reasons to be informed about the "latest advances," and now there are political and social ones as well. Knowing some of the facts and issues surrounding genetic testing or human cloning allows one to be part, and feel part, of the decision-making process. Some of this knowledge—including impressions—has come from media reporting itself, past and present. But this does not mean that scientists are required to obey such images in every instance.

What does the public need to know about a particular branch of scientific work? There is no simple answer. Indeed, the question itself is often misinterpreted. Researchers, that is, can all too readily confuse public *understanding* of science with public *appreciation*. Understanding (e.g., how nuclear energy is generated) can lead to queries, to criticism, and even to rejection. To know something of science is not necessarily to love it: the truly aware researcher must realize this and be prepared for it.

Interest in science is also affected deeply by medium. Consider that the great majority of media publications are meant to be skimmed, not studied: readers are able to retain very little specific information from a newspaper article, magazine story, or (especially) television or radio broadcast. This is due both to the style of exposure (quick, one-time reading or listening) and to the fact that there are usually many such exposures on a wide range of subjects (politics, international events, economics, features, etc.) to be ingested in a single sitting. The topical press is not something that provides people with opportunities for concentrated learning or continuing education. Reporters know this; they know they must write stories, not primers. Such is the other side of the coin from their need to engage.

Public interest in science is complex, multifaceted, difficult to define in any precise way. But one thing can be said for sure: in the popular press, this interest always comes back, sooner or later, to "news"—with everything this encompasses. In large part, it includes only the tip of the scientific pyramid—that part of science which is today in progress, being conducted, in the here and now. This is the most provisional—debated, competitive, uncertain—and, in a social sense, exciting part of science. But it is also the most difficult to write about in any definitive way. It is one thing to review for a public audience the basic principles of chemistry, which can be taken from the established literature; it is quite another to discuss the merits, hotly debated, of a new hypothesis on the physical chemistry of superconducting materials.

For scientists, it helps to distinguish between what is involved in knowing and knowing *about* a particular subject. Reporters write, and readers expect, stories *about* scientific ideas, advances, effects, and so forth. Technical knowledge per se is not the goal; the aim instead is to communicate something of the character, importance, and implications of this knowledge, what it might signify to the individual reader or viewer. This does not in the least excuse inaccurate reporting, sensationalism, or other common faults of the media. But it does help put boundaries around what scientists might anticipate from press coverage and public interest.

Being Prepared: Some Recommendations

What is the scientist's primary role vis-à-vis the press and the public? Most often, acting as an "expert source." What does this mean? In general terms, it means speaking as an inside witness to your own work, and very often to the state of knowledge, effort, and debate in your specific area. Your main duty, in other words, is to communicate the central points of your research, the history and limits of your knowledge, and what it might mean for your own field. Oh yes, and to add some human details too, about yourself, for instance.

Put this way, it all sounds pretty straightforward. But it can be intimidating to the unprepared scientist, who may be unsure about what to expect, what sort of tone to adopt, or how much to "reveal." If you work for an institution with a public relations office, especially one with a science officer or media consultant on staff, it is a good idea—no, a necessity—to meet with this person and discuss strategy before giving an actual interview, generating a press release, serving on a public panel, or the like.

If your research or expertise make it likely that you will be asked for an interview at some point, there are some things you can do to prepare. I don't mean hiring a publicist who is also a lawyer. There are much cheaper ways, such as these:

- Keep an eye out for media coverage related to your area of work. Note the sorts of questions being addressed, the type of language reporters use, the kinds of stories they tell, how the science is being handled. What do the writers get right and what do they get wrong (or less right), what strengths and weaknesses do you see?
- When you come across an excellent article, report, or documentary, save it for study. What do you like about it; why is it excellent? Is there anything you can adopt for discussing your own work?
- An *extremely* helpful groundwork for communicating with nonscientists is a concise (one-page), well thought-out summary of your work and its implications, in plain language. What would you say, given 30 seconds on national television or radio? Take time to write this carefully, then, when you're reasonably satisfied, read it over regularly, modify it as needed, show it to colleagues for comment; keep it fresh in your mind. It will help you be articulate, more confident, and, fortunately or no, a subject for future use.
- Creating such a summary, a valuable exercise on its own, will also give you sound-bite capability—an essential skill.

Reporters are neither your friend nor a prosecuting attorney, but they are very interested in cooperative contact. As a source, you are an asset to them, and may be so again in the future. Journalists would much prefer to discover and cultivate a new spring than poison the well. For the sake of producing an acceptable (and informative) article, they need to ask questions, some of which may appear to be critical or challenging.

These questions may be given to you in an email, for you to answer in text, or the reporter may want to interview you in person, either over the phone or face-to-face. If you feel comfortable answering or discussing the following list of queries, you should be in good shape to deal with most anything that might come your way:

- How did you come to be interested in this topic? Are there any personal reasons in your background that led you to it?
- Are there other researchers working on the same thing? Why do they think it is important?

- Why should readers care about your work? (the "So what?" challenge)
- How and where did you do your research? What methods did you use? How orthodox or unorthodox were these methods?
- Are your results reproducible? How consistent have they been between different studies?
- What alternative explanations might be given for your data?
- How much debate or controversy is there within your field? Who disagrees with you, and why?
- If your ideas prove correct, how might you profit from them, personally and professionally? (Indirect inquiries may try to get at this information.)

All of these are the kinds of questions that may come up, for example, in an interview or a series of ongoing discussions with a journalist. You should answer each one of them honestly, without any suspicions. Unless you're a highly controversial public figure, after all, and the reporter is too, she or he will want to avoid making you feel as if you're dancing through a minefield.

The In-Person Interview

Interviews are opportunities. Speaking or writing in a lucid way about your work or a related subject can do great service to you, your career, your colleagues, your institution, and your discipline. Here are some tips to help make sure this happens:

- First, make sure the reporter tells you what he or she is specifically calling about and what the deadline is, so you know whether he or she is in a rush or has time to listen to details.
- Second, be sure you are in a position to answer the reporter's questions. If the topic is related to a paper you haven't yet published, you should consider saying that it is too early to discuss your data. If you've submitted the paper to a journal, be sure you know its policy about talking to the press. Many journals have embargo policies that specify when your comments on the relevant research can appear.
- Get an idea of how much science the reporter knows, how much "homework" he or she has done on the topic (ask a question like "Did you know that . . . ?" or "Have you had a chance to read . . . ?"). This will tell you how basic your answers to questions might need to be. Two of the most com-

mon errors scientists make in interviews are to condescend to reporters and to use too much jargon.

- Reporters usually contact you ahead of time to schedule the interview for your convenience. Take the time to do two things: (1) write down the major points you'd like to cover (the reporter will only have space for a few of these, remember); and (2) think about how to talk about these points in especially clear and straightforward, jargon-free language.
- Google the reporter before the interview and look at other articles he or she has written. How good are they? Do they have any serious problems? If so, what are they? What might you say to prevent such difficulties?
- If the reporter wants to talk right away, ask to call him or her back in a little while (you can provide an excuse, if needed, for example, you have someone in your office). This will give you time to gather yourself and your thoughts.
- Avoid speaking on subjects outside your area of expertise. Scientists are sometimes treated as spokespeople on topics like political events, social movements, other disciplines (e.g., physicists holding forth on evolution, history). Don't take the bait. You'll appear far more honest and dignified if you say "I'm not really qualified to speak on that" or "That's not something I have expertise in" than if you spit out statements making you look ignorant or foolish, especially to real professionals in these areas.
- To the degree possible, don't pander to the general tendency of the press to speak in terms of "miracles," "revolutions," "perils," "magic bullets," and so on. Such language most often provides a public disservice in terms of characterizing the actual significance of a finding, hypothesis, or conclusion.
- Do not expect to see the final article before it is published. Newspapers and television stations (and their relatives) are businesses and do their own internal review. All the more reason, therefore, that you think before speaking. If some part of a story is truly in error, call the reporter and tell him or her (nicely) how best to correct it for the online version that remains in perpetuity.

Other Tips

Remember that your dealings with the press involve human relationships. To express open suspicion and distrust toward reporters is to effectively burn connections you might otherwise profit from.

- Be prepared to be asked about the social implications of your work. Reporters often want scientists to predict what the future might look like. Be clever; offer several scenarios. This will make you look especially thoughtful.
- If there is controversy or debate about the topic at hand, acknowledge this up front. Cast it in a positive light, in the sense that "critical review and scrutiny of new ideas are how science progresses."
- Be ready to point out the limits to your knowledge or that of your field. In this regard, try to avoid the "we don't know" type of response. In most cases, you are being consulted as a type of expert; therefore, speak in positive terms if you can, for example, about "questions now being pursued," "new areas of work opening up."
- Last, but by no means least, remember that, as a source, you cannot dictate in any final way what will be printed or said. It may help in this regard to reflect on the following:

> For the scientist who wants to be a source of science news, forewarned is forearmed: collaborating with journalists and adapting to journalistic conventions may give scientists more, rather than less, control over the emphasis and tone of the resulting story. But the last word will always go to the journalists, because science journalism is much more about journalism than it is about science.[5]

What If My Research Area Is Especially Controversial?

If you work in a subject area touched or tortured by public controversy, special conditions apply. Speaking in an interview or other media forum about topics like animal research, GMOs (genetically modified organisms), fetal research, and climate change often involves effort under pressure, even being put under a hot light. It may put you, that is, in the position of answering questions that are only partly scientific and meant to be provocative. For example,

- As an evolutionary biologist, how do you respond to the latest polling data that shows over half of all Americans think Darwin was a fraud?

---

5. J. Gregory and S. Miller, *Science in Public: Communication, Culture, and Credibility* (New York: Plenum, 1998), 130.

- What, in your opinion, accounts for the intensity of opposition to GMOs, even when they have been repeatedly tested and been shown to be safe?
- Do you think it is fair to use the word "alarmist" for some of the scenarios that climate scientists talk about regarding the end of this century?

Scientists have had to deal with these sorts of queries for a number of decades now, and only a few have become good at it.

If you view the situation as a chance to defeat the malevolent forces of ignorance or overturn what Carl Sagan once called "the demon-haunted world," you should reign your steed and reconsider. One of the worst things a researcher can do to his or her image and credibility, as well as that of science itself, is to appear like an arrogant crusader, condescending to the ignorati. Nuclear scientists found this out the hard way in the 1970s, when trying to deal with protests and public fears. An attitude of superiority (elite possession of the truth) will confirm the view that you don't really care what the public thinks and therefore aren't trustworthy. The surest way to alienate any audience is to act as if it doesn't matter.

Many scientists understand this today, of course. But the point can be lost in the heat of aggressive questioning. Reacting to such inquiry with phrases like "look, people need to realize that" or "it's easy to be led astray and to think that" or "no intelligent person can look at the data and claim" makes you an instant authoritarian. It is not advisable. Neither is calling an antiscience position like creationism a "refuge for those with the brains of table salt."

Controversies about science are emotional beasts. For the opposition, they are not only about lack of knowledge or bias or special interest, but about deeply held beliefs, ideas of how the world works (or should or doesn't work), and, of course, fear (many kinds). They therefore can evoke intense passions. Trying to win the argument, or even lower the tone, by use of facts alone will not work. Facts are important, without doubt, but they do not speak the same language as what I have just described, and, in truth, appear weak beside it, if not fortified with other elements.

When listening to an expert, the public is only too aware of its ignorance. Saying things that heighten this awareness isn't tactful. Moreover, those who are even the least bit skeptical or uncertain about scientific claims—and this can be a large part of any audience when a controversial topic is on the boil—are primed for the possibility that scientists will reveal themselves as part of an elite, closed-minded group. The key, then, is not to act accordingly. Instead, think of the audience as desiring new knowledge or perspective. This becomes more difficult when you know that many listeners or readers already have forceful opinions, including antiscientific ones. So what to do?

For starters, you have to keep your cool. This is evident, but its meaning isn't: remaining under control renders you a source of dignity and professionalism, which adds greatly to your authority. Keeping cool, however, doesn't involve going rigid like a robot and speaking without emotion. One approach, in an interview, is to remember that you're actually having a conversation. How do you talk to intelligent people? Your rhetoric is less defensive, more patient, at times enthusiastic, at others thoughtful. Instead of using the kind of phrases noted above, you might say something like "That's a good/excellent question. Let me try and answer it this way" (show respect for the reporter *and* the audience, then smoothly transition to take control of the discussion). Or, "Well, I'm not sure that's the best way to frame the question. What if we ask it this way (a courteous approach to refuting your interlocutor, while again taking control). Or, "Yes, certainly, there have been statements/concerns of that kind. And yet, the data we now have strongly point/don't allow us to agree" (give the appearance of value to the other side, then move to demolish it with scientific evidence). These same kinds of statements work excellently in a debate situation.

Some may feel this kind of language is too soft and gives too much credit to the opposing side, where no credit is due. Why can't we be assertive, call a spade what it is, state that the anti-Darwinians or climate change deniers are just dead wrong and have religious and political agendas? Because this plunges us into a rhetorical trap. It is a very old trap, in fact. Cicero once explained it like this: the best way to persuade an audience that your opponent has a valid point of view is to declare that he or she has nothing worthwhile to offer and should be prohibited from speaking. In Shakespeare's more well-known words: "The lady doth protest too much, methinks." By essentially calling for the censure of intelligent design or anti-GMO activism, you will make these viewpoints seem the victims of an elitist science. Far better, then, to keep control in your hands, avoid an angry or defensive posture, and seek to argue (kill) more softly.

As pointed out by scholars who study these controversies,[6] one of the most effective strategies opponents use to advance their claim is to con-

6.   See, e.g., L. Ceccarelli, "Scientific Controversy: Science, Rhetoric, and Public Debate," *Rhetoric and Public Affairs* 14, no. 2 (Summer 2011): 195–228.

tend that the science isn't settled. If an audience can be made to believe or even suspect that scientific debate remains open on a subject like evolution, then the antiscience position gains standing. Aiding the strategy is a view many people have of how science works. This is a romantic view that major advances often come from the work of a small number of heroic dissidents, whose unconventional ideas are first rejected and suppressed by the ruling orthodoxy. Its narrative can play well for the antiscience position when it adopts the language of an unfairly victimized minority.

In such cases, an effective strategy can be to emphasize the role that debate plays in modern science. This role very seldom involves a few against the many, but instead an extended discussion within the community of researchers that takes place through peer-reviewed publications. It is how new evidence, hypotheses, and ideas are clarified, challenged, tested, retested, and, finally confirmed or rejected by the community based on the best information from multiple sources. Such debate is central in what many understand as "the scientific method." It can be messy and may last for decades. But it is *a means to an end*, which is consensus about the most probable scientific truth. There are many active debates in science today—for example, whether humans were mainly responsible for the extinction of mammoths and other megafauna in the late Pleistocene—but there are a great many more that have been settled. This doesn't mean everyone agrees; geoscientists who rejected plate tectonics in the 1960s continued to argue and publish into the 1980s, well after the theory was established. But the mass of evidence in favor of the theory, and (just as important) the ever-growing span of phenomena the theory was able to explain, eventually made the dissenters vestigial. To reopen the debate would require oceans of credible evidence (claims alone can't be accepted). Regarding evolution, anthropogenic climate change, the safety of GMOs, the viral origin of HIV, the nonlink between vaccines and autism, debates have taken place and are scientifically settled. Arguments that claim the science remains unsettled are false and are not made for scientific reasons.

So let us return to the three questions with which we began this section. What might be a good response to each of them?

- Well, I would first say that I don't think Americans believe Darwin was a *fraud*, which in science means a cheat or a thief. Darwin was certainly neither of *those*. Some may think he was *wrong*, which is different. To that, I feel it's important to point out that the scientific debate about it has been settled for some time, just as it has been about gravity. Scientists today accept evolution as part of how nature works. Of course, this can still be rejected, but it would have to be on some other grounds than science.

- Yes, this is an important question. Fundamentally, I don't think it's really about scientific knowledge. The established facts about GMOs, proven by repeated testing in many countries, don't appear to count for very much. So the opposition seems to be about other things—distrust of corporations, for example, and their uses of science for profit without any ethical accountability. I think we need to talk about those things, keeping in mind that food, because we put it in our bodies, understandably attracts a lot of concern.

- Unfortunately no, I don't. I certainly wish it were otherwise, but every major domain of evidence we now have, both in the physical and the life sciences, points in a worrisome direction. Now, we have chosen to talk about the future in terms of scenarios—that is, different possibilities based on how much carbon we put into the atmosphere. The worst of these scenarios are indeed very alarming for the damage and suffering they would cause. But not "alarmist," because they do *not* exaggerate the case.

19. SCIENCE WRITING AND SCIENCE TALKS: COMMUNICATING WITH AND FOR THE PUBLIC

In the theater of the public's interest, the scholar
must have a speaking part.—CARL SAGAN

Don't Run—We Are Your Friends

The public would like to hear from you, the researcher. Whether or not you are convinced, it is patently true. Scientists are no longer an alien species to the mainstream, although the inverse cannot so easily be confirmed. Still, what might be called the "interest market" for science-related stories is large, and getting larger. The new century has made sure of this. How? Our lives are now filled with an array of technologies that continue to advance right before our eyes. On the dark side, there have been expanded worries about health and health care, diet and obesity; the possibility of pandemics and natural disasters (earthquakes, tsunamis); and the growing effects of climate change. On the positive side, we have seen spectacular developments: the decoding of the human genome, detection of the Higgs boson, exploration of the Martian surface, among others.

There is yet another reason for growing public interest. I have spoken of it several times already in this book. The Internet, particularly social media, has greatly reduced the space-time gap once dividing researchers from the rest of society. Websites, blogs, Facebook, Twitter, and other media all provide pathways for connection. As noted in the last chapter, this reality comes with real opportunity. People spend far more time online than reading hard copy newspapers and magazines, so having a sizeable following for your blog or tweets can be worth a good deal more than a single nice article mentioning your work, even in the *New York Times*. While never guaranteed, benefits mentioned by more than a few researchers have included increased citation of their work, invitations to give public lectures (i.e., chances to increase your influence further), and even improved funding.

In short, by writing in plain language about science, including your own, you can become a valued friend to the public. And in being such a friend, you may find portions of the public can requite your "affection" in valuable ways. The challenge, then, resides in your ability to communicate as needed.

What Is Science Writing?

Science writing is far easier to describe than define. It is definitely not *scientific* writing, which is what most of this book has been about. So? If we must have a definition of some kind, let it be no more difficult than "writing about science using ordinary, nonspecialist language." Now we can proceed to the real difficulties.

It is a myth, believed by some famous people, that everything in science can be expressed in plain language. It can't, and to say it can is foolish, or worse, misleading, because it suggests that specialist languages are both inessential and immoral, their real purpose being to prevent access. But while jargon does cleave the world into insiders and outsiders, this is not at all its epistemological purpose. Nor can it be called a waste of syllables. Trying to explain quantum mechanics or protein synthesis in everyday words will prove this—even these two titles, if defined fully, would require many pages and some level of simplification. The point is that the science writer has to be quite selective in what he or she includes.

This kind of writing, in other words, is not science per se. It is writing *about* science. Even if the topic is your own work, you are describing it, not presenting it to your peers. You are, in a sense, an expert-reporter. Such a position has its benefits and its challenges. The benefits should be obvious; in most cases, you'll fully understand what you're writing about and know where to get any needed sources. As for the challenges, the first one is major: you need to be able to write well. Purely functional writing will not be enough. Second, you must be partly a translator. The best science writing employs no jargon at all *unless* it is explained. This includes even terms like "DNA sequencing" or "natural climate variability," which appear often in media reports. Third, you must be able to tell your reader why the topic is both interesting and important, not so much for you but for them.

Good science writing, in other words, is a way to inform and teach. Put this way, it is far more than journalism normally conceived. You are taking an elite, expert knowledge—one with considerable authority and often great relevance to people's lives and futures—and making it available to the public. Moreover, this "public" is no longer restricted to a particular area, nation, or even continent. It includes, potentially, a true global readership, meaning anyone who is literate in the language you're using (especially if this language is English).

Put this way, science writing may sound like a noble endeavor. But this is precisely the case. It is all the more the case when scientists themselves are the yeomen in the fields of the word. By engaging in science writing, researchers avoid relying on an intermediary and, in a way, make themselves accountable. The opportunity can be weakened, even squandered, if the writer sees this type of writing chiefly as a vehicle for self-promotion. Most readers of science writing, however, are not fools and will disregard and avoid such stuff. Better, therefore, to put yourself forward as a quality spokesperson for your field. Such is an achievement, after all, of no small importance.

The How-To: Some Basics

Models of Good Writing: How to Choose Them

Science writing is a skill, like writing in general. The most direct and assured way to learn it is from those who do it well—good models, once again. Identifying such models—in newspapers, science magazines, blogs by other researchers, science websites, and so on—forms a first step. As with scientific texts, the feeling that "I'd like to have (I wish I had) written that" can be an indicator of worthy choices and can also be used to help determine the best among any group of models you may have chosen. Then comes the close study and rereading of them, so that they become your private mentors (of a sort) providing you with sources for emulation.

Fine, but how do you know when a piece of science writing that you like is, in fact, good enough to be such a mentor? What elements should you look for to judge a text's quality? We can list these in the form of questions to be answered.

- Is it *interesting*? Is the first sentence or opening paragraph likely to create interest in the topic for the average reader, and does the main body of the piece satisfy this interest?
- Is it *accurate*? Are the facts and numbers correct and are the concepts well-defined and properly used. Also, do any analogies and metaphors employed by the writer help add to the reader's understanding or not?
- Is it *easy to read*? Is the language simple and jargon-free? Can you pretty much read it from beginning to end without any clumsy or confused parts?
- Are there helpful *transitions* that move the text along, so that one thing follows another in a logical way to the very end?

These are not difficult questions, surely, but they come with a particular challenge of their own. In order to answer them usefully, you have to understand that you are not the audience (let me repeat, you are *not* the targeted reader). The audience is made up of people who know or remember little science, may well be intimidated by it, but are still willing to learn and be informed about it *if* it seems interesting enough and somehow relevant. Such people constitute the public for science writing. Your metamorphosis into this genus is absolutely essential. Few of your evaluations will be reliable otherwise.

Let me be more specific. Journalists maintain that the average reading level of US adults is that of an eighth grader. This view isn't intended to insult eighth graders everywhere, though it can hardly be said to compliment American adulthood. If this assertion makes your jaw drop and your eyes roll, think about it another way. Einstein said of science that "everything should be made as simple as possible, but no simpler." Applied to the case of the scientist who wants to do science writing, this maxim insists that much value can be derived from the discipline needed to make scientific material clear and reader friendly to eighth graders of all ages. If you have spent time lecturing and teaching, you grasp this immediately: simplifying knowledge down to its core concepts (but no further) demands the deepest understanding of the relevant science.

Techniques and Examples

As for the writing process, there do exist approaches and techniques that professionals regularly follow and talk about in some of their writings. First and foremost is the major emphasis placed on the opening of a text, what is called the lead. In science writing, the first few lines achieve two primary goals: they deliver the main take-home message and they create interest in the reader to read what follows. Here are two examples of published articles on the same topic:

A newly discovered extinct human species may be the most primitive unearthed yet, with a brain about the size of an orange. But despite its small brain size, the early human performed ritual burials of its dead, researchers say.[1]

A huge haul of bones found in a small, dark chamber at the back of a cave in South Africa may be the remnants of a new species of ancient human relative. Explorers discovered the bones after squeezing through a fissure high in the rear wall of the Rising Star cave, 50km from Johannesburg, before descending a long, narrow chute to the chamber floor 40 meters beneath the surface.[2]

We see, at first, clear differences between the samples. Their authors have not chosen the same information to present. Brains and burials are given in the first, bones and location in the second. In the first, we are met with a paradox: a brain no bigger than an orange may have invented religion. The second, meanwhile, offers a narrative: explorers penetrating the rear wall of a lightless cave, pressing on through a narrow fissure and chute. If alert, we note a problem of accuracy: is the new species "human" or a "human relative"? The metaphors deserve a look too. Using an orange to image a brain works pretty well (except for the color); everyone knows what an orange is, so a brain this size can be easily pictured. Nor is there any difficulty with "huge haul of bones" (note the alliteration) in the second example.

But what do the samples have in common? Their most fundamental aspects. Language level is almost identical. Both contain the same core message in their first sentence—the finding of a new species of hominid. Both samples also do a good job of creating interest, drawing someone's attention to what comes next. Notice, too, that each beginning sets up a series

1. Charles Q. Choi, "New Human Species with Orange-Sized Brain Discovered," *Live-Science*, September 10, 2015, http://www.livescience.com/52132-new-human-species-with -orange-size-brain-performed-ritual-burials.html.
2. Ian Sample, "*Homo naledi*: New Species of Ancient Human Discovered, Claim Scientists," *The Guardian*, September 10, 2015, http://www.theguardian.com/science/2015/sep /10/new-species-of-ancient-human-discovered-claim-scientists.

of questions for the reader that will need to be answered later on. In the first sample, these include how the discovery was made, why this new species is so primitive, whether its skull and body are also very small, and what the evidence for ritual burial is. For the second sample, the reader wants to learn the rest of the story about how the bones were found, retrieved, and identified. In both cases, an engaged reader will want to know the name of this new species and what it signifies, as well as where or whether the species fits into the existing ancestral tree.

Thus, in each instance, a large part of the article body is set up by the garden of questions that grow from these first lines. This is just what good science writing does. It does *not* spill information onto the page, like a debris flow. There is a real coherence to the whole that returns to the takeaway message of the beginning.

Now compare all of this with the beginning of the scientific paper on which these articles were based:

> Fossil hominins were first recognized in the Dinaledi Chamber in the Rising Star cave system in October 2013. During a relatively short excavation, our team recovered an extensive collection of 1550 hominin specimens, representing nearly every element of the skeleton multiple times, including many complete elements and morphologically informative fragments, some in articulation, as well as smaller fragments many of which could be refit into more complete elements.[3]

In this case, nonscientist readers would have questions mainly about the meaning of particular terms—"hominin," "skeletal element," "morphological," "articulation," "refit." More importantly, they wouldn't know that a new species had been discovered; this fact only comes toward the end of the paper, after all the evidence has been carefully described, analyzed, and considered. In an important way, science writing inverts the structure of a scientific paper, giving the final conclusion first.

This is as it should be. Creating real interest in readers at the beginning is the only way to draw them in so that they will spend time on the rest of an article. There's nothing cheap or scheming about such an approach.

3.   L. R. Berger, J. H. Darryl, J. de Ruiter, S. E. Churchill, P. Schmid, L. K. Delezene, T. L. Kivell, et al., "*Homo naledi*, a New Species of the Genus *Homo* from the Dinaledi Chamber, South Africa." *ELife* 4 (2015): e09560, http://elifesciences.org/content/4/e09560.

Nearly all writers employ it, from business and nonfiction authors to poets, novelists, and dramatists, including Shakespeare (the first words of *Hamlet*: "Who's there?"). No matter what level of education readers may have, the lead must do real work in getting them to read on. For another example, which of the following openings does this work better?

Since the accident at Fukushima, much new doubt has been expressed worldwide about the future of nuclear power. This has been especially true for the United States and nations in Europe.

In the immediate wake of Fukushima, many nations declared they would scale back plans for building new reactors. Europe and the United States made the strongest statements along these lines, not Japan.

Obviously the second opening is more interesting. Note how the first one merely gives us information. Its tone is flat, without emotion or suggestion, aided in this by the passive tense. We come away convinced that the rest of the article is no different and just fills in details, informing us but not teaching or engaging our minds.

The second sample offers an entirely different experience. Using the active tense, it portrays nations responding directly and fearfully to the Fukushima disaster. These nations are not just expressing doubt; they are doing something concrete: cutting back on future new builds. That introductory phrase, moreover, with the word "immediate," hints that they might actually have later reversed this decision, for reasons we hope to discover by continuing to read. Then there is the irony, or paradox if you like, that bursts out at the end with those two final words, "not Japan" (not present in the first sample). We are left wondering why the strongest antinuclear statements didn't come from the country where the disaster happened. Why *did* they emerge in countries on the other side of the world?

A good lead thus fulfills one more task besides generating interest and presenting the chief message. A good lead actually sets up the rest of the article (or blog post, op-ed, etc.), or a large part of it, by posing questions that need to be answered. Now it usually happens that leads which satisfy the first two tasks commonly do this, too—such flows from the success in presenting a key message in an interesting way.

Good leads come in many shapes and sizes. As we've seen already, some offer knew knowledge, some begin a narrative (story), while others work through irony or a quality of surprise. Still others combine two or more of these approaches into a paragraph that brings a degree of depth as well as stimulates curiosity. Here's an excellent example:

> A few years ago, Gene Robinson of Urbana, Illinois, asked some associates in southern Mexico to help him kidnap some 1,000 newborns. For their victims they chose bees. Half were European honeybees, *Apis mellifera ligustica*, the sweet-tempered kind most beekeepers raise. The other half were *ligustica*'s genetically close cousins, *Apis mellifera scutellata*, the African strain better known as killer bees. Though the two subspecies are nearly indistinguishable, the latter defend territory far more aggressively. Kick a European honeybee hive and perhaps a hundred bees will attack you. Kick a killer bee hive and you may suffer a thousand stings or more. Two thousand will kill you.[4]

A great deal might be said about this paragraph and how it plays on a reader. Having read the earlier part of this section, however, you should be able at this point to understand its workings without any guidance (so we can move ahead).

Writing a good lead sets you on an excellent path for what comes after. A simple approach is to deal with the questions posed by your lead in order. This may leave out important material if followed strictly, however. In the case of one or more experiments, you need to talk about what was done and why. If an expedition is involved, as in the first two samples above, a discussion is required of how and why it was launched, who funded it, where it went, and, of course, what it found. Beyond these kinds of details, or perhaps using them to advantage, you must answer the overriding "So what?" question: Why should the reader keep reading? What's in it for him or her? Good science writers rarely leave the answer unstated, unless it approaches the painfully obvious (a new cancer treatment, for example). But this doesn't mean the answer has to touch on a source of immediate concern or anxiety. Knowing that our human ancestry, thus our idea of the "human" itself, has been expanded by a new species with unanticipated features affects every one of us. Decisions about whether to grow

---

4. David Dobbs, "The Social Life of Genes," *Pacific Standard*, September 3, 2013, http://www.psmag.com/books-and-culture/the-social-life-of-genes-64616. I owe the use of this paragraph to science writer Carl Zimmer, who included Dobbs's article in a class he once taught.

or reduce the world's reliance on nuclear power have a number of big-time implications regarding climate change, coal use, energy security, and non-proliferation. And so on.

The main body of your text needs logic to make it flow and cohere. Look for an order of connection among your major points. There is always one to be found or forged, unless your material is too scattered and disparate. A host of phrases awaits your use for this purpose: "that being said"; "what this implied was"; "the next thing to consider"; "we weren't convinced, suspecting that"; "physicists have long accepted this"; "her next step was to" Questions can help establish or move forward the logic in a text, too: "Why hadn't we seen this before?" "How long had they been there?" "What kind of process could possibly be at work?" "Where did that result come from?" "Can it be true that . . . ?"

If you are writing a blog post, you can certainly let your voice carry a bit more into the lands of opinion. Ideas about the social, political, or economic relevance of your topic, for example, that would ordinarily be viewed as too speculative for a newspaper or magazine article, are absolutely fine here, as long as you tell your readers that they are in fact your views. Indeed, such is one of the advantages to blogging—you have no censors, no editors, no external authorities with needs and motives that differ from your own.

We finally come to the ending, where techniques also matter. The idea here is not to end in a boring or nondescript manner, whether with lifeless phrasing or in textual drool. Readers should be left with something pointed, thoughtful, even powerful. As an example, take the sample above dealing with the Fukushima nuclear accident. This is its finale:

> A better future for humanity depends on expanding the reach of modern energy while reducing its carbon footprint. Lowering the use of coal and oil will not be easy, quick, or cheap. But it can be greatly advanced by the nuclear option, designed to the highest standards and managed to the same. Such is a truth now understood by a growing number of nations, who have come to accept even in the aftermath of Fukushima that climate change is a more real and certain threat to the human prospect.

Thus, a dramatic kind of ending. A different approach closes the example about the bees, an article that is actually about social behavior and genes. It ends with a quote from a key researcher, who ends his public lectures this way: "Your experience today will influence the molecular composition of your body for the next two to three months," he tells his audience, "or, perhaps, for the rest of your life. Plan your day accordingly."

The Public Talk: Practical Advice from the Front Lines

We looked briefly at some main points about public talks by scientists in chapter 13, where the context was oral presentations. Here, I will amplify and add to these in the frame of communicating with nonscientists. Those who do not live under rocks or in labs far underground know that this subject deserves this kind of varied coverage because of the rising importance such communication has in today's world.

If speaking to nonscientific audiences is something that interests you or will likely be asked of you, keep in mind three overriding requirements: first, you must be a human being, an interesting one if at all possible; second, you must be accurate in what you say, that is, tell the truth (about what is known and what isn't; about issues, debates, controversies); and third, if your subject is scientific knowledge, you must remain a scientist and not slip into the role of advocate or activist.

Public talks, like all speeches, are a type of performance. To be good, they must be stimulating to hear *and* watch. People do not come to witness data. Nor do they desire to be preached or pandered to. They want a living, breathing scientist who has done something or knows something important, even exciting, and can tell them about it in a way that preserves its truth and gives it a pulse.

Perhaps I'm telling you that snow is white. But perhaps not. It is worth noting that the entire rhetorical tradition reaching back to classical Greek oratory lays a good deal of stress and force on the influence that comes from eloquent speeches eloquently delivered. And while some did employ the techniques of eloquence in a mercenary way, for causes just or unjust, the greatest speeches of that time, such as Demosthenes's *On the Crown*, were those that "embraced truth" and avoided empty persuasion in the name of a personal belief or political promotion. Hollow eloquence, however, is not very often the problem with talks by scientists. How many talks have you sat through where the researcher qua person just didn't show up: no anecdotes, no personal details, no emotion or enthusiasm, no humor (intentional, that is), no real contact with the audience—in other words, no performance. Just a specimen-scientist exhibiting slides. Can we doubt that far too many of these talks occur on a daily basis? They may satisfy the second two requirements mentioned above—accuracy and nonadvocacy—but they fail in the first and, in some ways, the most important.

So consider giving a talk that you yourself (as a scientist from an unrelated field) would enjoy and find memorable. What kinds of things would grab your interest and own it? Wouldn't they involve some of the elements just mentioned above—meaningful stories, expressions of enthusiasm, bits of humor, and so on? Probably they would, as long as the science itself were accurate and the main point.

How, then, to design and deliver a talk that, short of chiseling your name among the immortals of oratory, will engage and even inspire your listeners? Guides on public speaking are numerous. This isn't surprising, since such guides have been around for more than two millennia in the West. Possibly the most famous is the *Rhetorica ad Herennium* (literally "Rhetoric for Herennius"), written by an unknown author around 80 BCE. This remained a core textbook on eloquence for over 1,600 years, with editions still being published as late as the 19th century. Calling it highly influential is an understatement; it served as a key text in the history of Western education. Its primary aim was to be useful for political and juridical speaking, but it was found applicable to any situation where the speaker wanted to persuade an audience to his or her point of view. Even today, the *Ad Herennium* provides a flexible template for this purpose, which is why, knowingly or not, most modern guides follow in its footsteps. The template has six parts:[5]

- Introduction (*exordium*), where the speaker prepares the listener's mind to be "well-disposed" and receptive, using the promise of "important, new, and unusual matters, or such as appertain to the . . . hearers themselves." Meaningful anecdotes, quotes from famous authors, or analogies (e.g., for science, "Launching this kind of project was like . . .") are also helpful.
- Narration (*narratio*), where the speaker concisely describes the main thrust of the talk, whether this involves a discovery or thesis or summary; some mention is also given to the people involved and something about them and their "traits of character," as if setting the stage for a story.
- Division (*divisio*), where the speaker lays out "briefly and completely,

5. The quotations that follow are from [Cicero], *Rhetorica ad Herennium*, trans. Harry Caplan, Loeb Classical Library (Cambridge, MA: Harvard University Press, 1954).

the points we intend to discuss" (no more than three) and might also mention what makes them unique, how the result of any work or thought differs from what others have claimed or supported about the subject.

- Proof (*confirmatio*), where the speaker gives "the presentation of our arguments, together with their corroboration," that is, evidence. Incorrect interpretations of the evidence may also be presented, as a way to create a kind of staged debate that can help keep the audience fully engaged.
- Refutation (*refutatio*), where such opposing arguments are refuted, along with any possible objections to the speaker's ideas.
- Conclusion (*conclusio*), where the speaker offers a concise summary of the argument, along with its implications.

If we put all these parts together, in the form of a single progression or arc, what we see is a speech or talk that at every step has the audience's response directly in view. Where the introduction is meant to establish a friendly, even warm connection, the narration and division take a bit of time and space to prepare and ready the listeners for the principal substance of the talk, the proof. The refutation, then, even if quite short, shows that you have considered other ideas, interpretations, views, as a good scholar must—a potent sign to the audience of your legitimacy and commitment to the truth. Finally, as a summary, the conclusion demonstrates that you want your hearers to actually carry something away from their experience, that they do learn something new.

What seems nice about the six-part delivery is that it lays out a map for the level of emotion/enthusiasm that might go into a talk: warm at the beginning, somewhat more serious into the narration and division, more eager in the proof, lighter in the refutation, and, at the last, calm and confident in the conclusion. You can envision how the speed of your discourse might change accordingly: slower at the beginning, faster in the middle, slower again toward the end.

None of this is very far at all from the type of advice you'll find in the guides of today. Like the *Ad Herennium*, these will also do their best to impress upon you the value of practice. No substitute exists for this; you need to give the talk a number of times until you become comfortable with it, until you refine it to where it no longer feels like a burden (terrible or otherwise) but something you really do want to share. Yes, it is possible to practice too much, to overtrain, so that the material grows stale ("the laborious is not necessarily the excellent," says the *Ad Herennium*). It is always good to go on stage with a little edge. It helps feed your enthusiasm.

As for language, when you talk to a nonscientific audience, it makes sense to combine informal, colloquial discourse with somewhat more formal speech in your most serious points. Being too informal when discussing the core science of your talk can have a trivializing effect; try to avoid phrases, metaphors, and analogies that make you sound as if you are pandering to a less intelligent audience (case in point: the use of pop culture or sports metaphors).

Many public speaking guides will tell you it is essential to first decide exactly what message you want to convey, then to carefully plot out how you're going to convey it, filling in details one by one. No doubt this works in some cases, depending on the subject. But it should never be considered as a universal rule, to be strictly obeyed. While it can certainly help some scientists, it would mean creative death for others. As with writing, important points and ideas emerge in the process of building a talk. While it nearly always helps to create an outline at the beginning, filling this in with details and either writing a script or producing slides (or both) will inevitably suggest changes, even big ones. Remaining open to change can prove essential. Good talks may reveal their own internal logic as they mature in process; connections and secondary themes emerge.

Some scientists I know who give excellent talks employ intuition or trial and error in building a lecture for public audiences. They begin with a striking image or a brief story that they know will grab attention, then look for a following image or a transition to the science that moves deeper into the topic. If their talk is heavy on visual imagery, as the center of attention, they may speak without notes, almost extemporaneously, discussing each image in turn. Other experienced speakers, however, have told me they work by dividing a talk into sections, listing main points (including anecdotes or other spices) under each, and then writing a script in outline form, even before starting on any slides.

What follows is a list of speaking tips that come from a wide range of sources, modern and ancient, as well as from my own experience and my discussions with scientists whose talks I have admired (and borrowed from). Some you will find in other guides, but a few you probably won't. For easier and more selective use, I have divided them according to different aspects of speaking.

Delivery
- Address your talk to those in the back row—they must be able to hear you, and if they do, everyone else will too.

- Don't turn and talk to the screen; if you are pointing at something on a slide, turn halfway for a moment to show what it is, then, while discussing it, turn back to the audience again. They are the focus of all your attention.
- Therefore, talk *to* the audience, not at them; look at them directly, though not always in the same place. Pick out some faces in different parts of the audience to speak to, as if personally (if they are more than about 20 feet away, it will seem you are looking at a number of people, not only them). This use of your own attentiveness shows real interest in the people before you, giving them the feeling that you really want them to understand this very cool subject, which you do, or you shouldn't be there in the first place.
- Smile, at least occasionally (the topic is something you also enjoy working on, right?). This creates warmth and human connection and will make the audience much more forgiving if you happen (that one-in-a-million chance) to misspeak or stumble.
- Changes in pace can be very effective. First, speak a bit more slowly than usual overall. Use pauses now and then to let points sink in. As suggested above, try to vary the speed of your speech now and then; for example, when something is exciting, speed up a little and increase the volume. One researcher I know would sometimes speak too quickly when he got to the parts of his talk where he discussed his own work; he would then stop, apologize, and say something like "This just means a great deal to me and my colleagues." This had a powerful effect on people, evoking their empathy. Remember that a monotone voice (the "lawnmower") is the executioner of a willing audience.
- Try to look and sound confident; stand up straight, keep your arms fairly open, so that you are not hunched forward or closed up (more open posture denotes confidence, command, and welcome).

Content
- Please don't begin by introducing yourself and the topic of your talk or saying "thank you for inviting me" and other "blah-blah" openings. The audience knows basically who you are and what you'll be talking about, having been informed by a poster, via email, or some other way. Starting in any of these utterly unexceptional ways will tell people you have nothing unique to say and may well be blessed with the talent to numb their brains into inactivity.

- Good ways to begin:
  - With a question that will almost certainly touch on something that interests the audience. For example, "Will we ever be able to cure cancer?" or "What do we *really* know about the origins of human beings?" or "What role can chemistry have in helping us create a new kind of vehicle?"
  - With a striking fact (or factoid) related to your topic. For example, "There are now roughly three trillion trees in the world today" (for a talk on forestry, habitat loss, climate change) or "Every 40 minutes, the amount of energy reaching the Earth's surface from the Sun equals the total energy consumed by all of humanity in an entire year." Such facts can be looked up easily and, with only a little practice, used effectively. They can be used in a clumsy way, too (the "Hey kids, guess what?" approach).
  - With a narrative opening, a "once upon a time" for adults, that promises to embed the science in a tale of the people behind it, along with crucial events. A simple way to do this is to begin with the words: "I want to tell you a story" or "I'd like to take you on a journey" or the like. Corny in writing, but when spoken with a smile and a touch of eagerness, this kind of opening will make the audience yours. We have all, as children, been trained to love stories told in a caring voice. Follow through by talking about the people who did the science, especially if it was you and your colleagues. Who are they, why did they choose such a topic, what did they sacrifice, what dead ends did they face—in short, the human details that were involved in the creation of new scientific knowledge.
- Certain verbal cues and transitions can help move your talk along at key points: "so this is what we did"; "now, this part is important"; "if we look very close here, we can see"; "what we found was not at all what we expected to find"; "so now we understood that."
- Use of questions can be very effective at keeping an audience's focus: "But what did *this* mean?" "Why did they decide to . . . ?" "Was there another explanation?"
- If you are using slides, you must make them visually interesting; shift out of your role as a communicator to other scientists in your discipline; you should be without mercy in reducing words to an absolute minimum in plain language (the eighth graders in the audience will love you for this).
- As with scientific presentations, you can help defeat the evil demon of

stage fright by writing out the first several lines or paragraphs of your talk and memorizing them, so you can launch your talk with confidence.

- Quotes from famous thinkers, including scientists, can be worthy to include, but they need to be considered carefully. Al Gore's use of them for "An Inconvenient Truth" was famously effective because of careful choice and strategic use (Mark Twain, for thoughtful emphasis). Please do not pander to your audience with clichés, such as the "wonders of science" and "mysteries of the universe." Beware, too, of relying on such *expected* sources as Einstein, Watson and Crick, and Darwin. Quotations, then, shouldn't just be cute or sassy filler, but should actually *do* something—emphasize a specific point, summarize a key section, pose a central question, help you shift to a new theme.

- Enthusiasm is more important than you may think; in fact, it is *extremely* important. People come to a talk not just to learn something new but to be stimulated into doing so; they have given up other activities to be here. You achieve something profound by showing them they made the right choice. You can do this quite easily by complimenting your material: "we're going to be talking about some truly exciting/awesome/fascinating things"; "what comes next surprised everyone"; "this is great stuff here"; and so on. Obviously this can be overdone; but if you say such things only two or three times in a 45-minute talk, it will be quite effective.

- Keep in mind that your audience wants most to hear the positive side of the science you're discussing. They don't wish to hear you complain about or judge other researchers and their work. Temporary failures that eventually led to successes make excellent stories; failures by themselves do not.

- Finally, be wary of how you portray science itself. As noted elsewhere in this book, the public tends to have a fairly primitive view of how science works, a view unfortunately supported by media reports. This view sees the great majority of scientists contributing small, plodding advances, with major progress coming from true geniuses and from maverick outsiders who challenge the given order, suffer calumny, but are finally proven right. Such ideas deserve to be changed, not confirmed.

Practice

- Rehearsing your talk is not an option but an absolute necessity, like speaking in a language you know. It is only by speaking—not going through your talk in your head or watching videos of excellent oratory—that you become a better speaker. You need to become fully aware of how you sound.

- Therefore, start practicing from a fairly early stage, when you have about a third of the talk done. This will show you which parts are easy to move through and which parts require more effort or some kind of change.

- Be aware of how quickly you are speaking. Identify those places where you tend to be too slow or fast, where you hesitate or struggle a little (or a lot), and use this awareness to adjust, to smooth things out.

- A good way to practice is to record yourself and watch the result. Most, if not all, laptops, tablets, and smartphones now have cameras that make doing this extremely easy. Don't worry about your appearance in these; focus on your speech, expressions, and gestures. Do this kind of self-recording and evaluation several times at least before rehearsing in front of a colleague or other friend.

Last Words

Researchers today who wish to give public talks have a definite advantage over their predecessors. It's called YouTube. With it, scientists are able to find good and even excellent talks that can be used as models (partial or whole) and watched as many times as desired, for free. They can even search for "best science talks" and be introduced to a growing reservoir of possible choices. It is certainly worth checking some of these out.

More than a few of the most highly rated talks are from the TED (Technology, Entertainment, Design) conferences (https://www.ted.com). These talks, then, deserve some discussion. While a number of TED science talks are indeed admirable and would serve very well as models, it's important to keep in mind that they are designed for a special kind of venue and audience. These talks are rarely more than 20 minutes long. The filmed versions focus heavily on the speaker, showing few of the slides that might have been used. This is great for the speaking side of things, but it won't help with the graphical side. Talks take place on a theater-type stage, with the speaker as the overriding visual focus, something that differs significantly from what a scientist will usually face. Some other positives: less-experienced speakers are often heavily coached and their talks scripted to hold viewer interest well; they often include anecdotes, humor, and quotes, and speaker enthusiasm is nearly always on display. Talks have strong openings and good pacing, too. Only one other aspect might be pointed out with caution, and that is the audience. This is highly selective, including only invited individuals, most of whom are highly educated and all of whom have paid $6,000 (the ticket price in 2015). Thus a TED talk is not entirely suited for an average audience, yet this will depend on the talk.

I would suggest going through some of these online, finding several you especially like, and studying them to see how they work and what may need to be adjusted for your own audience(s). Different scientists will inevitably have their own favorites; some of my own include Carolyn Porco's "This Is Saturn," Brian Cox's "CERN's Supercollider," David Gallo's "Underwater Astonishments," and Bonnie Bassler's "How Bacteria 'Talk.'" It is worth spending time to find those you think might best aid your own public presentations. But remain open to other talks on the Internet, as there is much beyond the terrain of TED. As always, choose your mentors wisely since their influence may remain with you for some time.

20. TEACHING SCIENCE COMMUNICATION: HELPFUL IDEAS FOR THE CLASSROOM

Those who can, do; those who know,
teach. —IRENE SCHWARTZ (teacher)

Acknowledgments

My experiences in teaching students and career scientists how to improve their communication skills have been nothing if not instructional for the instructor. No class is ever the same. Each of them is a new challenge. I have taught seminars with five or six people, medium-sized groups of 15–25, and even large clusters of 40–50. These classes have been held in universities, in government labs, at corporate headquarters, at conferences and other professional meetings (various scientific fields), and in living rooms, for native and nonnative English speakers, and for instructors of writing too.

Through the forest, patterns of light have emerged with time. Teaching people how to better communicate involves a degree of risk. I refer not to a risk of rejection or disappointment. What I mean is the possible failure to change students in lasting ways. The intensely personal nature of communication, be it writing or speaking, requires that students (whoever they may be) alter their sense of who they are and what they are capable of, as they accept new competence. All true learning is about identity; thus, the responsibilities of teaching are wide and deep and the triumphs always uncertain.

My point is this: changing people's ability to communicate often begins with changing their sensibility about communicating. Such is how this book began—much begins with attitude. Here, in this final chapter, it makes sense to talk about how this might be done with a rewarding degree of success.

The reader will note that I am not offering a detailed template or tablet of commands for how to teach a class, day by day, week by week. Teachers do not like to be told how to conduct their classes, especially according to formula, or if they do, then something might be wrong beyond what I can hope to address here. In any case, what follows is a series of suggestions and ideas that I have found, over time and with varied experience, highly useful to guide my design and teaching of successful courses on writing and speaking.

Teaching Scientific Writing and Speaking

In this section I provide some how-to ideas that I have found work well in all of the teaching situations I've encountered in two decades of doing such work. These ideas are not at all unique to myself and whatever insight I might possess but have come from much reading, learning from excellent instructors, and (forgive the terminology) clinical classroom trials. While it is certainly possible to learn effective teaching techniques and put them to use in an organized, rational way, such an approach by itself will no more make your skill and sense of achievement as a teacher highly rewarding than an impeccable knowledge of grammar will make you a novelist.

Teaching is about the creation of experiences. And there is a reason why this is especially true in the case of writing and speaking. It is mainly because of the personal dimension and the sense of exposure (vulnerability) that go along with communicating to strangers, including those in a classroom. As a result, it can help greatly for a teacher to establish a relationship of emotional engagement and trust with students. This is no less true if the class is composed of midcareer professionals or undergraduates. How to do this?

Try to learn everyone's name, as best you can. I can't overstate the importance of this, as it creates a real connection (you are *recognizing* each one of them). See if there is a class list, or, best of all, a class list with photos (many universities now provide this to faculty). If nothing is available, ask people their names as they respond to questions in class—a better approach than going through at the beginning and asking each person to sound off his or her name in military fashion, one after the other. Learn a memory technique for keeping the names straight, and when you make a mistake, joke about it to relieve any tension ("Sorry about that—I'm still recovering from a major skull fracture").

Stop and tell stories now and then, whether about your own experience as a teacher, an author, a speaker, about something that happened in another class you taught, or about something you read or heard that applies to writing and speaking. These stories should be meaningful and entertaining, too ("I had this editor once who was always angry, angry at the language or what his writers were doing to it, and he never let a single sentence of mine escape without injury"). These stories can be true, but they don't really have to be if their point is accurate and worthwhile. They are teaching tools. At least some of them should really be about you, and here, of course, a bit of self-deprecation is far better than any kind of bragging ("This reminds me of one rejection I had, where the review comments said just about everything except that I was depriving some poor village of its rightful idiot").

As the figure of authority and judgment, you are granted many powers. Because of this, a few kind words go a long way. Complimenting students when they give a thoughtful or right answer to a question can do a great deal for engagement and motivation. Simple phrases, such as "nice idea," "that's well said," "good point," "I hadn't thought of that," and "this is an impressive class," may do wonders for morale and thus effort. At the same time, the opposite will happen if such comments are reserved for a small minority of students who are especially outgoing or overconfident. Encouraging words help in such a case, for example, "How about a few new voices on this question?" And, if nothing results, "Okay, I've seen some good work from a number of people, how about you, Marsha, what are your thoughts on this?" Meanwhile, if having something of a stern exterior (and perhaps interior) is your thing, as the sign of professionalism and dignity, consider smiling now and then to warm the atmosphere.

Use some of your own writing when doing analyses of published samples. I've found that using such samples without identifying you're the author until afterward works best. Also, it can help if you first provide lesser-quality samples (and tell the class these were early efforts), then superior samples later on, thus showing your own advancement, whether this be during a single class or over a week or more.

In designing a course, as this guide suggests, a worthwhile approach is to divide the overall plan into two basic parts, the first devoted to reading, the second to writing. The correlative for talks would be a first part on listening followed by a second part on speaking. I concentrate on reading and writing here.

The first step would be to help students learn how to read critically. This means how to analyze a piece of scientific writing and evaluate it. Such analysis can be done in a number of different ways—by sections of the paper, by rhetorical function, by flow, logic, and transitions, among others—and there is no reason to recommend one type of analysis over another. I have offered one such type in my discussion of Watson and Crick's famous DNA paper (see chapter 2).

A key goal is for students to come away with a clear set of questions to ask of any text. Or, to put it another way, they should have a concrete idea of what to look for in any scientific publication. For example, Does each section of the paper fulfill its role? Does the writing flow smoothly and logically? Are there good transitions, so that it doesn't jump around from point to point? Is there an introduction and a conclusion? Is the terminology well-chosen (that is, are terms used correctly, or is there too much jargon)? Such are the kinds of questions that help evaluate the quality of the writing. On top of this, of course, there are content queries about the science. For example, Are the methods acceptable? Is the data convincing? Are the conclusions well enough supported? And so on.

A good idea is to have students discover some of the writing-related questions themselves by giving them poor or lesser-quality papers to read and comment on. Doing this verbally in class works well when you have students willing to volunteer and speak up; it also gives you a chance to provide compliments. But if the class is less forthcoming, this can be done silently and turned in, or provided as homework and gone over in class.

The second part of the course or workshop will then focus on writing exercises. To make direct use of the critical reading skills already learned, students can be given a paragraph, again from a weak paper, to first critique and then rewrite. These rewrites can then be used as further class material if students partner with each other, exchange their versions, and perform critiques on them.

There are many kinds of writing exercises or projects that can be done, certainly. I'll say only that students do need to see examples of high-quality writing, to analyze what makes them good, and to be given the chance to emulate them. This is far easier to do if your class lasts a week or longer. For a one- or two-day workshop, this may not be possible, though it can be figured into the work if examples are kept quite short.

Teaching Science Writing and Speaking

Some teachers know that running a course or workshop for scientists who want to learn to write for the public can be enjoyable, even fun. It can also be excruciating. Pleasing when your students are motivated to learn a new skill, such a class becomes insufferable when they either doubt the value of any such skill or feel impossibly superior to it. I have had the experience of both types of classes, many times.

The reader will not find it surprising that nearly everything said in the first part above applies here as well, with some important additions and subtractions. A helpful way to begin, or something to do at an early stage, is to contrast science writing and scientific writing. Much has been said about the topic, and this book provides a number of major points about the differences (see chapter 19). An effective approach, and one that is often used, is to compare an original scientific paper with one or more popular articles reporting on its findings. Placing the opening paragraph(s) of each side by side offers a clear demonstration that students can observe and comment on.

Whether you have to deal with resistance regarding the value of science writing or not (you may not know at first, of course), it helps to go over the advantages anyway. It is certainly fine to talk about access to knowledge in a democracy (or in *any* kind of nation-state) and the obligation of scientists to government funding, thus the public. But you may find it also worthwhile to point out and discuss the benefits this skill of science writing can have for a researcher's career, namely, the expanded exposure of his or her work and consequent influence and the chance to improve public knowledge and serve as a positive representative for science itself (no small thing).

In analyzing a piece of science writing, the things to look for are different than in a scientific paper. Some questions that might be asked include, Is it interesting; does it try to engage us or just spill information onto the page? Can it be read easily, by a nonscientist (no unexplained jargon)? Is the information accurate? Do any metaphors or analogies used by the author work well with the subject (or do they seem odd, out of place, or strained; e.g., "the resulting tsunami erupted into the coast")? Does the text flow logically, with good transitions and internal connections? Does it have a strong ending?

Another approach is to ask what story is being told here. This helps focus the perspective on people and events and allows the instructor to trace the arc of the narrative in an article. Depending on the size and makeup of a class, it may make sense to begin this way and then focus on the questions given above.

In terms of finding and using good models, you (the instructor) can supply these well enough, or you can have the students search for some that they especially like. A helpful guide for them to search with is the "I'd like to have written that!" criterion (or, among several choices, "I'd *really* like to be able to write something like that").

If everyone in the class is from the same scientific discipline, then the choice of models by yourself or other students can be held up to evaluation by everyone. In other words, the class (with some guidance, if needed) can debate and finally decide which examples they would most like to emulate. On the other hand, if you have scientists from several disciplines, it is more difficult to do this. In this case, I would suggest not sticking to examples from just one field.

Some time spent on the title is a good idea. Everyone recognizes the importance here, that titles are the first thing anyone sees and therefore most often determine whether a reader will go any further. Again, showing poor, mediocre, and good examples is worthwhile. So is having students rewrite some poorer samples and then comparing their versions in class.

Finally, a goal to possibly strive for in any class is having students reach a level of comfort such that they are willing to volunteer their own writing (and rewrite examples) for critique by others. In my experience, only a few people will ever do this early on or in a one- or two-day course. You may be lucky and find a more courageous group under your tutelage or have a method for making this happen very soon (my congratulations if you do; please do tell me about it). Otherwise, it becomes a real mark of success if a class reaches the point where its members have the confidence to subject their work to in situ peer review.

In the end, the teaching of communication skills may appear different from other types of courses in higher education. In fact, it is different. It falls somewhere between teaching a humanities subject like literature and a skill-based trade such as translation. This makes it challenging, risky, and hugely rewarding when successful. Teachers (I am one) count themselves among the benefactors of society, as they should, when someone leaves their embrace a different, more knowledgeable and skillful person than when he or she entered it. It is to aid this possibility that I have offered the humble volume of advice in this chapter. I have learned a fair portion of it from other teachers, such as Irene Schwartz (see the opening quote to this chapter), once upon a time an exceptional eighth-grade instructor of English, whose unfathomable patience and caring expertise certainly changed the life of one unruly student.

21. IN CONCLUSION

We often mistakenly think that education is
something we can finish.—ISAAC ASIMOV

I began this book with the perception that research and communication form a continuum. In the end, it is more accurate to say that, in the real world of daily scientific work, they are inseparable. Research involves a number of central activities, and communicating is one of them. Using words is interwoven with laboratory work, library work, theoretical work, collegial contact, and every other type of labor associated with science, from beginning to end. If you can't communicate adequately with your peers, you can't do research, at least in any truly productive sense. Poor writing makes for bad science.

Such, at least, is how a stern parent might put the matter. I prefer to say, more gently (but no less realistically), that the sharing of knowledge—especially in formal, communal ways—is the nutritive process that makes the body of science a living, growing enterprise. Reading, speaking, and writing, with at least a functional level of skill, thus become the inevitable responsibility of every scientist who wishes to contribute directly to the vitality of that corpus. Presenting one's work to others therefore need not be an act of mere survival. It can be a conscious deepening of one's participation in a domain that, after all, one has chosen for a lifetime of effort and loyalty.

Contrast this with the standard idea that the scientist must "write up" his or her research. What is meant by this little phrase, so often used, so rarely questioned? Partly, it recalls the dreaded lab report of school science, with its demand to get required data down on paper before "writing up" the results (in apparent seesaw fashion). Something of this carries over into professional work, where this phrase suggests that the investigator must first call a halt to *real* work—in the lab, the field, the office, wherever—before beginning to write. He or she must then sit down, in monastic manner, draw breath, and wrestle the beast of language into submission. Writing thus appears an unfortunate, even lamentable obstacle, an intrusive obedience to outside demands.

Such attitudes, of course, are fateful. Shall we say destructive, self-fulfilling? Writing is work, certainly, just as any process of experimentation and discovery must be—but no harder than this, and no less significant.

Writing does involve a journey to the interior, and for many, this may not be an especially pleasant experience. There can be significant anxiety about giving expression to one's work, for, in truth, it is a type of exposure. Yet I wager that if we go back far enough in our own private histories, to when we first felt attracted to science as lifework, we will find embedded in our nascent and possibly cinematic images of "science" itself a certain desire to influence others, to compete and ascend, and to share and connect with a larger world to which we would add something important. The hope of such addition, and the ambitions it nucleated, rested on the sense of making our own mind and work available to others.

Like any author, scientists add themselves to the world through symbols; this is how they make their contribution. The scientist today has a greater, more exciting range of expressive outlets to employ than ever before. It is, indeed, an exciting and demanding era. There are more forms of publication to be aware of, more media to learn and take advantage of, more avenues of contact among professionals and institutions. Science is much larger than it once was, more urgent, diverse, complex, hungry. But communication remains its core substance. More than ever, scientists are scholars of the written word. Knowing how to write with a modicum of functional skill or better, and even a degree of pride or pleasure, gives one of the most essential endowments for negotiating successfully the demands of the scientific life.

It has been my effort in this book to outline ways that will help the student and practitioner of science acquire or advance this endowment. A principal theme has been the use of models of good writing as a basis for improving one's own expression. This is hardly an original idea; indeed, it is among the most ancient of wisdoms. But it requires time, and therefore patience. We continue to need good writers to keep our science vital and growing, and if we cannot train them fully during their school years, we need to provide methods for them to develop thereafter. Knowingly or not, all writers learn by example, and by experiment. This is a sine qua non of any art or craft. Making this process conscious, even methodical, can therefore grant one considerable advantage, whether as an apprentice or self-styled journeyman.

"The world is a noisy business," said Daniel Defoe, one of the most prolific authors who ever put pen to paper (more than 500 works may have come from his hand). If we are to count writing in general a major contribution to this noise, then certainly science is the source of untold reverberations that carry us all from the quieting past into a boisterous future. Modern science largely began as literature—the sharing of knowledge and experience through the publication of books, journals, diaries, translations, and more. Literary it may no longer be, but literature, in the larger sense, it certainly remains. The great library of science becomes greater with each passing year.

SELECTED BIBLIOGRAPHY

*Books on Scientific Writing, Illustration,
Editing, and Related Topics*

Alley, M. 1996. *The Craft of Scientific Writing*. 3rd ed. New York: Springer.

————. 2013. *The Craft of Scientific Presentations*. 2nd ed. New York: Springer.

Anholt, R. R. H. 1994. *Dazzle 'Em with Style: The Art of Oral Scientific Presentation*. New York: W. H. Freeman.

Baron, N. 2012. *Escape from the Ivory Tower: A Guide to Making Your Science Matter*. 2nd ed. Washington, DC: Island Press.

Bishop, C. T. 1984. *How to Edit a Scientific Journal*. Baltimore: Williams and Wilkins.

Blicq, R. S. 1995. *Writing Reports to Get Results: Quick, Effective Results Using the Pyramid Method*. 2nd ed. New York: Institute of Electrical and Electronics Engineers.

Blum, D., M. Knudson, and R. M. Henig, eds. 2006. *A Field Guide for Science Writers*. 2nd ed. New York: Oxford University Press.

Booth, V. 1993. *Communicating in Science*. 2nd ed. Cambridge: Cambridge University Press.

Borgman, C. 2010. *Scholarship in the Digital Age: Information, Infrastructure, and the Internet*. Cambridge, MA: MIT Press.

Borowick, J. N. 1996. *Technical Communication and Its Applications*. New York: Prentice Hall.

Briscoe, M. H. 1996. *Preparing Scientific Illustrations: A Guide to Better Posters, Presentations, and Publications*. 2nd ed. New York: Springer-Verlag.

Ceccarelli, L. 2011. "Scientific Controversy: Science, Rhetoric, and Public Debate." *Rhetoric and Public Affairs* 14 (2): 195–228.

Chambers, J. S. 1983. *Graphic Methods for Data Analysis*. London: Chapman and Hall.

Cleveland, W. S. 1993. *Visualizing Data*. Summit, NJ: Hobart Press.

————. 1994. *The Elements of Graphing Data*. Rev. ed. Boca Raton, FL: CRC Press.

Coghill, A. M., and L. R. Garson. 2006. *The ACS Style Guide: A Manual for Authors and Editors*. 3rd ed. New York: American Chemical Society.

Council of Science Editors. 2014. *Scientific Style and Format: The CSE Manual for Authors, Editors, and Publishers*. 8th ed. Chicago: University of Chicago Press.

Davis, M. 1997. *Scientific Papers and Presentations*. San Diego: Academic Press.

Day, R. A. 1995. *Scientific English: A Guide for Scientists and Other Professionals*. 2nd ed. Phoenix, AZ: Oryx Press.

Dean, C. 2009. *Am I Making Myself Clear? A Scientist's Guide to Talking to the Public*. Cambridge, MA: Harvard University Press.

Dear, P., ed. 1991. *The Literary Structure of Scientific Argument*. Philadelphia: University of Pennsylvania Press.

Duarte, N. 2008. *Slide:ology: The Art and Science of Creating Great Presentations*. Sebastopol, CA: O'Reilly Media.

Fourdrinier, S., and H. J. Tichy. 1988. *Effective Writing for Engineers, Managers, Scientists*. 2nd ed. New York: John Wiley & Sons.

Gant, S. 2007. *We're All Journalists Now: The Transformation of the Press and Reshaping of the Law in the Internet Age*. New York: Free Press.

Glasman-Diel, H. 2009. *Science Research Writing for Non-native Speakers of English*. London: Imperial College Press.

Goodlad, S. 1996. *Speaking Technically : A Handbook for Scientists, Engineers and Physicians on How to Improve Technical Presentations*. London: World Scientific Publishing Co.

Gopen, G. D., and J. A. Swan. 1990. "The Science of Scientific Writing." *American Scientist* 78 (6): 550–558.

Gregory, J., and S. Miller. 1998. *Science in Public: Communication, Culture, and Credibility*. New York: Plenum.

Hailman, J. P., and K. B. Strier. 2006. *Planning, Proposing, and Presenting Science Effectively: A Guide for Graduate Students and Researchers in the Behavioral Sciences and Biology*. 2nd ed. Cambridge: Cambridge University Press.

Hayes, R., and D. Grossman. 2006. *A Scientist's Guide to Talking with the Media: Practical Advice from the Union of Concerned Scientists*. New Brunswick, NJ: Rutgers University Press.

Hers, H.-G. 1984. "Making Science a Good Read." *Nature* 307 (5256): 205.

Hodges, E. R. S. 1988. *The Guild Handbook of Scientific Illustration*. New York: John Wiley and Sons.

Hoover, H. 1980. *Essentials for the Scientific and Technical Writer*. 2nd ed. Mineola, NY: Dover.

JAMA Archives. 2007. *American Medical Association Manual of Style : A Guide for Authors and Editors*. 10th ed. New York: Oxford University Press.

Kenny, P. 1982. *Public Speaking for Scientists & Engineers*. Bristol, UK: Adam Hilger.

Lannon, J. 1979. *Technical Writing*. Boston: Little, Brown, and Co.

Levine, G., ed. 1987. *One Science: Essays in Science and Literature*. Madison: University of Wisconsin Press.

Lindsay, D. 1995. *A Guide to Scientific Writing*. 2nd ed. New York: Longman.

Locke, D. 1992. *Science as Writing*. New Haven, CT: Yale University Press.

Matthews, J. R., J. M. Bowen, and R. W. Matthews. 1996. *Successful Scientific Writing: A Step-by-Step Guide for Biomedical Scientists*. London: Cambridge University Press.

Medawar, P. B. 1990. "Is the Scientific Paper a Fraud?" In *The Threat and the Glory*, 228–233. New York: HarperCollins.

Meredith, D. 2013 *Explaining Your Research: How to Reach Key Audiences to Advance Your Work*. New York: Oxford University Press.

Montgomery, S. L. 1996. *The Scientific Voice*. New York: Guilford.

———. 2000. *Science in Translation: Movements of Knowledge through Cultures and Time*. University of Chicago Press.

———. 2013. *Does Science Need a Global Language? English and the Future of Research*. Chicago: University of Chicago Press.

Morgan, S., and B. Whitener. 2006. *Speaking about Science: A Manual for Creating Clear Presentations*. Cambridge: Cambridge University Press.

Moriarty, M. F. 1997. *Writing Science through Critical Thinking*. Boston: Jones and Bartlett.

Nelkin, D. 1995. *Selling Science: How the Press Covers Science and Technology*. New York: W. H. Freeman.

O'Connor, M. 1991. *Writing Successfully in Science*. New York: Harper Collins.

Paradis, J. G., and M. L. Zimmerman. 1997. *The MIT Guide to Science and Engineering Communication*. Cambridge, MA: MIT Press.

Pechenik, J. A. 1993. *A Short Guide to Writing about Biology*. 2nd ed. New York: Harper Collins.

Scanlon, E., R. Hill, and K. Junker. 1999. *Communicating Science: Professional Contexts*. London: Routledge, in association with the Open University.

Scanlon, E., E. Whitelegg, and S. Yates. 1999. *Communicating Science: Contexts and Channels, Reader 2*. London: Routledge, in association with the Open University.

Schimel, J. 2012. *Writing Science: How to Write Papers That Get Cited and Proposals That Get Funded*. New York: Oxford University Press.

Schoenfeld, R. 1989. *The Chemist's English*. New York: John Wiley & Sons.

Shortland, M., and J. Gregory. 1991. *Communicating Science: A Handbook*. New York: John Wiley & Sons.

Sides, C. S. 1991. *How to Write and Present Technical Information*. Phoenix, AZ: Oryx Press.

Slade, C., and R. Perrin. 2007. *Form and Style: Research Papers, Reports, Theses*. 13th ed. Belmont, CA: Wadsworth Publishing.

Stapleton, P. 1987. *Writing Research Papers: An Easy Guide for Non-native English Speakers*. Canberra: Australian Centre for International Agricultural Research.

Suber, P. 2012. *Open Access*. Cambridge, MA: MIT Press.

Tufte, E. R. 1983. *The Visual Display of Quantitative Information*. Cheshire, CT: Graphics Press.

———. 1990. *Envisioning Information*. Cheshire, CT: Graphics Press.

———. 1997. *Visual Explanations*. Cheshire, CT: Graphics Press.

Valiela, I. 2000. *Doing Science: Design, Analysis, and Communication of Scientific Research*. Oxford: Oxford University Press.

Wilkinson, A. M. 1991. *The Scientist's Handbook for Writing Papers and Dissertations*. New York: Prentice Hall.

Williams, J. M. 1995. *Style: Toward Clarity and Grace*. Chicago Guides to Writing, Editing, and Publishing. Chicago: Chicago University Press.

Wolff, R. S., and L. Yeager. 1993. *Visualization of Natural Phenomena*. New York: Springer-Verlag.

Wood, P. 1994. *Scientific Illustration: A Guide to Biological, Zoological, and Medical Rendering Techniques, Design, Printing, and Display*. 2nd ed. New York: John Wiley and Sons.

Worsley, D., and B. Mayer. 1989. *The Art of Science Writing*. Philadelphia: Teachers & Writers Collaborative.

Yang, Jen Tsi. 1995. *An Outline of Scientific Writing for Researchers with English as a Foreign Language*. World Scientific Publishing Co.

Zimmerman, D. E., and D. G. Clark. 1987. *The Random House Guide to Technical and Scientific Communication*. Random House.

Zinsser, W. 1985. *On Writing Well: The Classic Guide to Writing Nonfiction*. 6th ed. New York: Harper Reference.

INDEX

Page numbers in italics refer to illustrations.